Edward Dobson

Rudiments of the Art of Building

In Five Sections

Edward Dobson

Rudiments of the Art of Building
In Five Sections

ISBN/EAN: 9783743400788

Manufactured in Europe, USA, Canada, Australia, Japa

Cover: Foto ©berggeist007 / pixelio.de

Manufactured and distributed by brebook publishing software (www.brebook.com)

Edward Dobson

Rudiments of the Art of Building

RUDIMENTS OF THE

ART OF BUILDING

In Five Sections—

GENERAL PRINCIPLES OF CONSTRUCTION; MATERIALS USED IN
BUILDING; STRENGTH OF MATERIALS; USE OF MATERIALS;
WORKING DRAWINGS, SPECIFICATIONS, AND ESTIMATES

By EDWARD DOBSON, A.-M.INST.C.E. AND M.I.B.A.,
AUTHOR OF "PIONEER ENGINEERING," "MASONRY AND STONECUTTING,"
"BRICKS AND TILES," ETC.

With numerous Illustrations

THIRTEENTH EDITION

LONDON
CROSBY LOCKWOOD AND SON
7, STATIONERS' HALL COURT, LUDGATE HILL
1890

PREFACE.

In offering this little volume to the public, it may be desirable to say a few words, by way of preface, as to the object and character of the work. It has been written at the suggestion of the publisher, to accompany the Rudimentary Series, and as a first book on the Art of Building, intended for the use of young persons who are about to commence their professional training for any pursuit connected with the erection of buildings ; and, also, for the use of amateurs who wish to obtain a general knowledge of the subject without devoting to it the time requisite for the study of the larger works that have been written on the different branches of construction.

To avoid unnecessarily extending the limits of the work, those subjects are omitted which are treated of in other volumes of this series, as Building Stone, Brick-making, and the Composition of Colours and Varnishes. For the same reason little has been said of the manufacture of glass and the smelting of metallic ores, because they have been repeatedly treated of in various elementary works, whilst a considerable space has been devoted to the consideration of the differences between hot and cold blast irons, and to the description of the operations of the iron-founder, subjects which are not generally to be met with but in expensive works.

The equilibrium of retaining walls is a subject which has long engaged the attention of mathematicians with little practical success, the results arrived at by different eminent writers being quite at variance with each other. For the chapter on this subject a few simple formulæ are given, which embrace all the conditions of the thrust of the earth and of the resistance of the wall, the friction of the earth against the back of the wall being also taken into account *.

In the article on the strength of cast-iron flanged beams a simple rule is given for calculation, founded on the assumption that the position of the neutral axis in a cast-iron rectangular beam, at the time of fracture, is at about ⅛th of its whole depth below its top surface, which is now pretty generally admitted to be the case. Amongst various works, the following have been carefully consulted during the composition of this little work The publications of the Institution of Civil Engineers and of the Royal Institute of British Architects, Professional Papers of the Royal Engineers, Weale's Quarterly Papers on Engineering and Architecture, Weale's Bridges, the Works of Peter Nicholson, Gwilt's Encyclopædia of Architecture, Dr. Ure's Dictionary of Arts and Manufactures, Tredgold's Carpentry, the works of Pasley and Vicat on Limes and Cements, Aikin's Papers on Arts and Manufactures, Barlow on the Strength of Materials, Tredgold and Hodgkinson on Cast Iron, and Bartholomew on Practical Specifications ; all these works will be found extremely valuable to the student.

I have great pleasure in acknowledging the kind assistance of my friend MR. H. W. KIRBY, C.E., in the articles on the equilibrium of retaining walls, and on the strength of cast-iron beams; to whom I am also indebted for the valuable notes appended to the article on retaining walls.

The articles on Iron-Founding, Carpenter and Joiner's Work, and House-Painting, have been carefully revised by friends practically engaged in those pursuits.

E. DOBSON.

CONTENTS.

SECTION II.

MATERIALS USED IN BUILDING.

SECTION III.

STRENGTH OF MATERIALS

SECTION IV.

USE OF MATERIALS.

SECTION V.

WORKING DRAWINGS, SPECIFICATIONS, ESTIMATES, AND CONTRACTS

LIST OF ILLUSTRATIONS.

SECTION I.

SECTION II.—No ILLUSTRATIONS.

SECTION III.

SECTION IV.

RUDIMENTS

OF THE

ART OF BUILDING.

SECTION I.

GENERAL PRINCIPLES OF CONSTRUCTION.

FOUNDATIONS.

1 In preparing the foundation for any building, there are two sources of failure which must be carefully guarded against: viz., inequality of settlement, and lateral escape of the supporting material; and, if these radical defects can be guarded against, there is scarcely any situation in which a good foundation may not be obtained.

2. *Natural Foundations.*—The best foundation is a *natural* one, such as a stratum of rock, or compact gravel. If circumstances prevent the work being commenced from the same level throughout, the ground must be carefully *benched out*, i.e. cut into horizontal steps, so that the courses may all be perfectly level. It must also be borne in mind that all work will settle, more or less, according to the perfection of the joints, and therefore in these cases it is best to bring up the foundations to a uniform level, with large blocks of stone, or with concrete, before commencing the superstructure, which would otherwise settle most over the deepest parts, on account of the greater number of mortar

B

joints, and thus cause unsightly fractures, as shown in fig. 1.

Fig. 1.

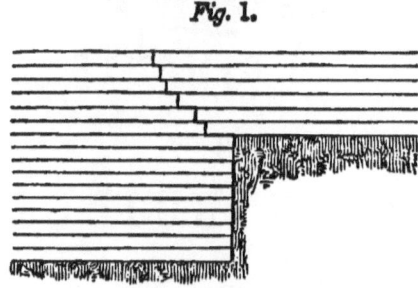

3. Many soils form excellent foundations when kept from the weather, which are worthless when this cannot be effected. Thus blue shale, which is often so hard when the ground is first opened as to require blasting with gun powder, will, after a few days' exposure, slake and run into sludge. In dealing with soils of this kind nothing is required but to keep them from the action of the atmosphere. This is best done by covering them with a layer of concrete, which is an artificial rock, made of sand and gravel, cemented with a small quantity of lime. For want of this precaution many buildings have been fractured from top to bottom by the expansion and contraction of their clay foundations during the alternations of drought and moisture, to which they have been exposed in successive seasons.

4. *Artificial Foundations* — Where the ground in its natural state is too soft to bear the weight of the proposed structure, recourse must be had to artificial means of support, and, in doing this, whatever mode of construction be adopted, the principle must always be that of extending the bearing surface as much as possible ; just in the same way, that, by placing a plank over a dangerous piece of ice, a couple of men can pass over a spot which would not bear the weight of a child. There are many ways of doing this —as by a thick layer of concrete, or by layers of planking, or by a network of timber, or these different methods may

be combined. The weight may also be distributed over the entire area of the foundation by inverted arches.

5. The use of timber is objectionable where it cannot be kept constantly wet, as alternations of dryness and moisture soon cause it to rot, and for this reason concrete is very extensively used in situations where timber would be liable to decay.

6. In the case of a foundation partly natural and partly artificial, the utmost care and circumspection are required to avoid unsightly fractures in the superstructure; and it cannot be too strongly impressed on the mind of the reader, that it is not an *unyielding*, but a *uniformly yielding* foundation that is required, and that it is not the *amount*, so much as the *inequality*, of settlement that does the mischief.

The second great principle which we laid down at the commencement of this section was—To prevent the lateral escape of the supporting material. This is especially necessary when building in running sand, or soft buttery clay, which would ooze out from below the work, and allow the superstructure to sink. In soils of this kind, in addition to protecting the surface with planking, concrete, or timber, the whole area of the foundation must be inclosed with piles driven close together;—this is called *sheet-piling*.

An example of a wide-spread foundation in soft ground is shown in fig. 2 (p. 4), which is a section of the foundation for the walls of the Leyden station of the Amsterdam and Rotterdam Railway, built A.D. 1843.* The station stands upon such bad ground, that it was necessary to support the walls upon a kind of raft resting on oak piles.

7. Where there is a hard stratum below the soft ground, but at too great a depth to allow of the solid work being brought up from it without greater expense than the circumstances of the case will allow, it is usual to drive down

---

* From the "Minutes of Proceedings of the Institution of Civil Engineers," 1844.

wooden piles, shod with iron, until their bottoms are firmly fixed in the hard ground. The upper ends of the piles are then cut off level, and covered with a platform of timber on which the work is built in the usual way.

Fig. 2.

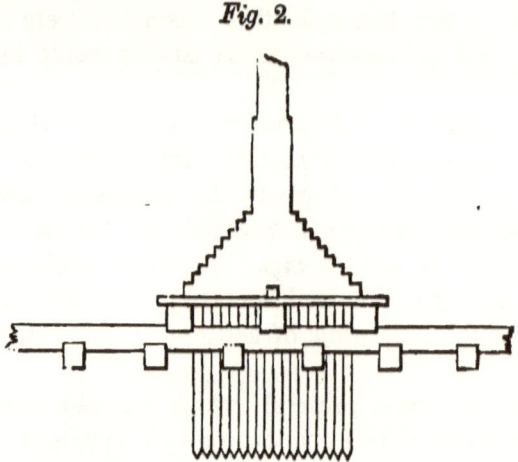

8. Where a firm foundation is required to be formed in a situation where no firm bottom can be found within an available depth, piles are driven, to consolidate the mass, a few feet apart over the whole area of the foundation, which is surrounded by a row of sheet-piling to prevent the escape of the soil; the space between the pile heads is then filled to the depth of several feet with stones or concrete, and the whole is covered with a timber platform, on which to commence the solid work.

9. *Foundations in Water.*—Hitherto we have been describing ordinary foundations; we now come to those cases in which water interferes with the operations of the builder, oftentimes causing no little trouble, anxiety, and expense.

Foundations in water may be divided under three heads: 1st, Foundations formed wholly with piles. 2nd, Solid foundations laid *on* the surface of the ground, either in its natural state, or roughly levelled by dredging. 3rdly, Solid foundations laid *below* the surface, the ground being laid dry by cofferdams.

10. *Foundations formed wholly of piles.*—The simplest foundations of this kind are those formed by rows of wooden piles braced together so as to form a skeleton pier for the support of horizontal beams; and this plan is often adopted in building jetties, piers of wooden bridges, and similar erections where the expense precludes the adoption of a more permanent mode of construction; an example of this kind is shown in fig. 3.

Fig. 8.

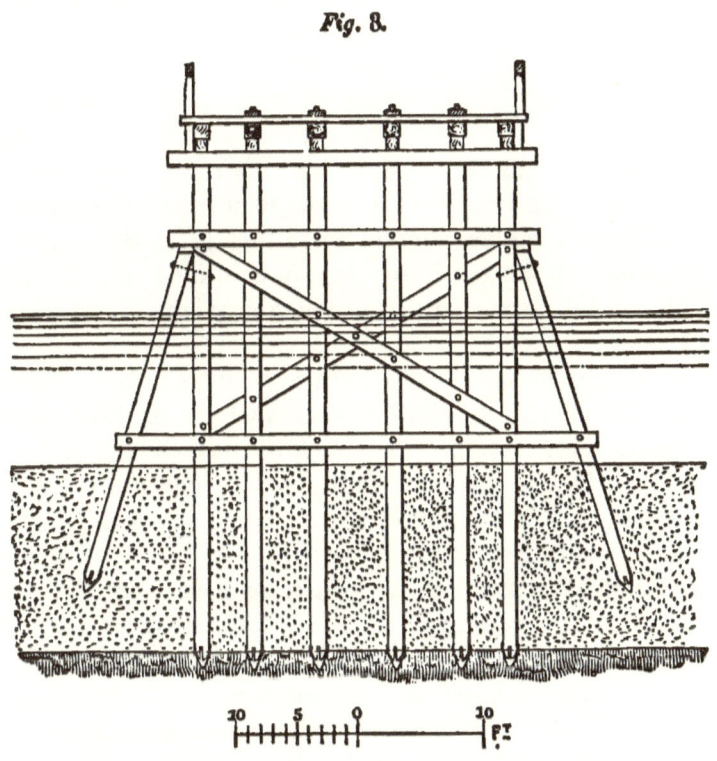

In deep water the bracing of the piles becomes a difficult matter, and an ingenious expedient for effecting this was made use of by Mr. Walker, in the erection of the Ouse Bridge, on the Leeds and Selby Railway, A.D. 1840. This consisted in rounding the piles to which the braces are at-tached for a portion of their length, to allow the cast-iron

sockets in which they rest to descend and take a solid bearing upon the square shoulders of the brace-piles. After the
brace-piles were driven, the braces were bolted into their
sockets and dropped down to their required position, and
their upper ends were then brought to their places and
bolted to the superstructure

11. There is always, however, a great objection to the use
of piles partly above and partly under water, namely, that,
from the alternations of dryness and moisture, they soon
decay at the water-line, and erections of timber require extensive repairs from this cause.   In tidal waters, too, they
are often rapidly destroyed by the worm, unless great expense is undergone in sheathing them with copper.

To obviate the inconveniences attending the use of
timber, cast iron is sometimes used as a material for piles:
but this again is objectionable in salt water, as the action of
the sea-water upon the iron converts it into a soft substance
which can be cut with a knife, resembling the Cumberland
lead used for pencils.

12. In situations where a firm hold cannot be obtained
for a pile of the ordinary shape, such as shifting sand,
Mitchell's patent screw-piles may be used with great advantage.   These piles terminate at the bottom in a large
iron screw 4 ft. in diameter, which, being screwed into the
ground, gives a firm foot-hold to the pile.   This is a very
simple and efficient mode of obtaining a foundation where
all other means would fail, and has been used in erecting
light-houses on sand-banks with great success.   The
Maplin sand light-house at the mouth of the Thames, and
the Fleetwood Lighthouse, at Fleetwood, in Lancashire,
both erected A.D. 1840, may be instanced.

13. An ingenious system of cast-iron piling was adopted
by Mr. Tierney Clark, in the erection of the Town Pier at
Gravesend, Kent, A.D. 1834, in forming a foundation for the
cast-iron columns supporting the superstructure of the T
head of the pier.   Under the site of each column were
driven three cast-iron piles, on which an adjusting plate was

firmly keyed, forming a broad base for the support of the column, which was adjusted to its correct position, and bolted down to the adjusting plate

14. A kind of foundation on the same principle as piling has been lately much used in situations where ordinary piling cannot be resorted to with advantage. The method referred to consists in sinking hollow cast-iron cylinders until a hard bottom is reached. The interior of the cylinder is then pumped dry, and filled up with concrete or some equally solid material, thus making it a solid pier on which to erect the superstructure. The cylinders are made in lengths, which are successively bolted together as each previous length is lowered, the excavation going on at the bottom, which is kept dry by pumping. It often happens, however, in sinking through sand, that the pressure of the water is so great as to blow up the sand at the bottom of the cylinder; and, when this is the case, the operation is carried on by means of a large auger, called a miser, which excavates and brings up the materials without the necessity of pumping out the water. The lower edge of the bottom length of each cylinder is made with a sharp edge, to enable it to penetrate the soil with greater ease, and to enter the hard bottom stratum on which the work is to rest. This method was adopted by Mr. Redman in the erection of the Terrace Pier at Gravesend, Kent, finished A.D. 1845.

15. Before closing our remarks on pile foundations, we must mention a very curious system of carrying up a foundation through loose wet sand, which is practised in India and China, and is strictly analogous to the sinking of cast-iron cylinders just described.

It consists in sinking a series of wells close together, which are afterwards arched over separately, and covered with a system of vaulting on which the superstructure is raised. The method of sinking these wells is to dig down, as far as practicable, without a lining of masonry, or until water is reached; a wooden curb is then placed at the bottom of the excavation, and a brick cylinder raised upon it

to the height of 3 or 4 ft. above the ground. As soon as
the work is sufficiently set, the curb and the superin.
cumbent brick-work are lowered by excavating the ground
under the sides of the curb, the peculiarity of the process
being that the well-sinker works under water, frequently
remaining submerged more than a minute at a time.
These cylinders have been occasionally sunk to a depth of
40 ft.

16. *Solid Foundations simply laid on the Surface of the
Ground.*—Where the site of the intended structure is perfectly firm, and there is no danger of the work being undermined by any scour, it will be sufficient to place the materials on the natural bottom, the inequalities of surface being
first removed by dredging or blasting.

17. *Pierre perdue.*—The simplest mode of proceeding is
to throw down masses of stone at random over the site of
the work until the mass reaches the surface of the water,
above which the work can be carried on in the usual manner. This is called a foundation of *"pierre perdue,"* or
random work, and is used for breakwaters, foundations of
sea-walls, and similar works. Plymouth breakwater is an
example on a large scale.

18. *Coursed Masonry.*—Another way, much used in harbour work, is to build up the work from the bottom (which
must be first roughly levelled) with large stones, carefully
lowered into their places; and this is a very successful
method where the stones are of sufficient size and weight
to enable the work to withstand the run of the sea. The
diving-bell affords a ready means of verifying the position of
each stone as it is lowered.

19. *Béton.*—On the Continent foundations under water
are frequently executed with blocks of béton or hydraulic
concrete, which has the property of setting under water.
The site of the work is first inclosed with a row of sheet
piling, which protects the béton from disturbance, until it
has set. This system is of very ancient date, being described by Vitruvius, and was practised by the Romans, who

have left us many examples of it on the coast of Italy. The French engineers have used béton in the works at Algiers, in large blocks of 324 cubic feet, which were floated out and allowed to drop into their places from slings. This method, which proved perfectly successful, was adopted in consequence of the smaller blocks first used being displaced and destroyed by the force of the sea

20. *Caissons.*—A caisson is a chest of timber, which is floated over the site of the work, and, being kept in its place by guide piles, is loaded with stone until it rests firmly on the ground. The masonry is then built on the bottom of the caisson, and when the work reaches the level of the water the sides of the caisson are removed.

This method of building has been much used on the Continent, but is not much practised in this country. Westminster Bridge, London, is a noted instance of its failure. The bottom of the river has been scoured out to a depth of several feet since the erection of the bridge; and the foundations of the piers remained in a dangerous state until they were secured in the recent repairs by driving sheet-piling all round them, and underpinning the portions which had been undermined.

21. An improvement on the above method consists in dredging out the ground to a considerable depth, and putting in a thick layer of béton on which to rest the bottom of the caisson.

22. There is a third method of applying caissons which is practised by our continental neighbours, and which is free from the objections which commonly attend the use of caissons. A firm foundation is first formed by driving piles a few feet apart over the whole site of the foundation. The tops of the piles are then sawn off under water, just enough above the ground to allow of their being all cut to the same level. The caisson is then floated over the piles, and, when in its proper position, is sunk upon them, being kept in its place by a few piles left standing above the others, the water being kept out of the caisson by a kind of well con-

structed round each of these internal guide piles, which are
built up into the masonry   This method of building in
caissons on pile foundations is shown in figs. 4 and 5.   The

Fig. 4.

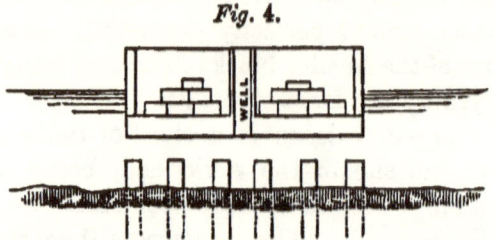

piers of the Pont du Val Benoît at Liége, built A.D. 1842
which carries the railway across the Meuse, have been built
on pile foundations in the manner here described.

Fig. 5.

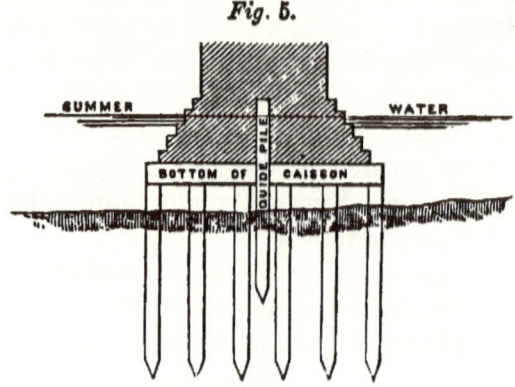

23. *Solid Foundations laid in Cofferdams.*—There are
many circumstances under which it becomes necessary to
lay the bottom dry before commencing operations.   This is
done by inclosing the site of the foundation with a water-
tight wall of timber, from within which the water can be
pumped out by steam power or otherwise.   Sometimes, in
shallow water, it is sufficient to drive a single row of piles
only, the outside being protected with clay, as shown in
fig. 6; but in deep water two or even four rows of piles will
be required, the space between them being filled in with
well-rammed *puddle*, so as to form a solid water-tight mass.

Fig. 6

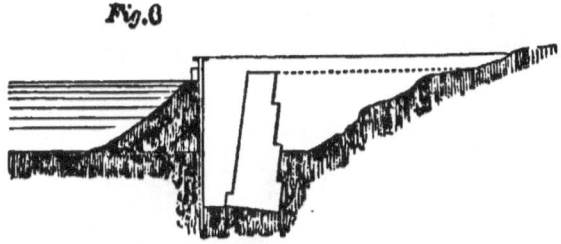

'See fig. 7) The great difficulties in the construction of a cofferdam are—1st, to keep it water-tight; and, 2nd, to support the sides against the pressure of the water outside, which in tidal waters is sometimes so great as to render it necessary to allow a dam to fill to prevent its being crushed

Fig. 7.

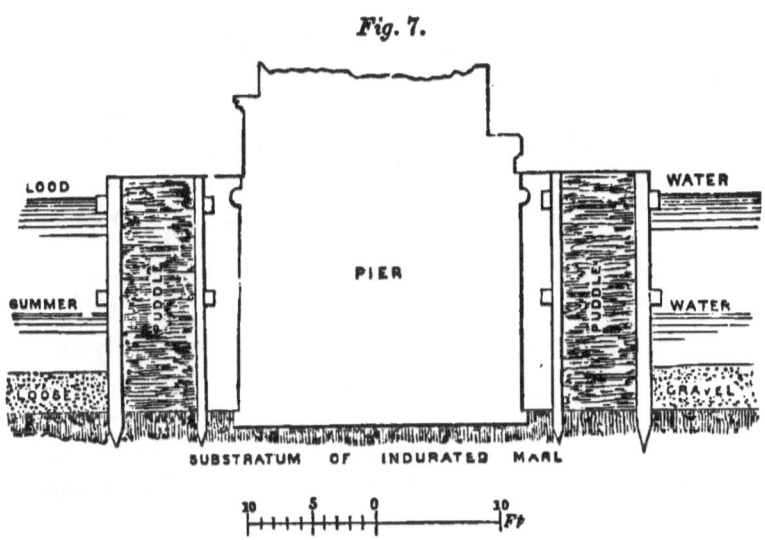

24. In order to save timber, and to avoid the difficulty of keeping out the bottom springs, it has been proposed by a French engineer, after driving the outer row, to dredge out the area thus inclosed, and fill it up to a certain height with béton. The cofferdam is then to be completed by driving an inner row of piles resting on the béton, and puddling between the two rows in the usual manner; and the

masonry is carried up on the béton foundation thus pre-pared. This construction is shown in fig. 8.

Fig. 8.

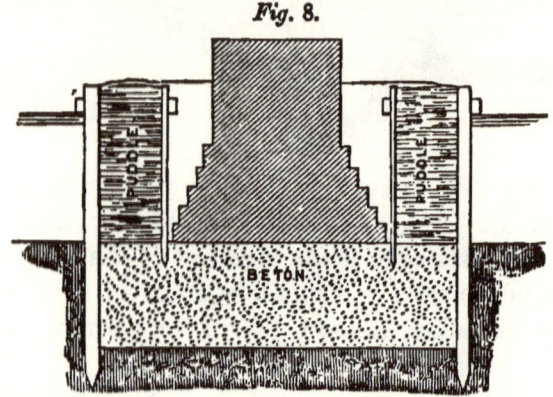

25. The limits of the present volume prevent our entering into any detail as to the preparation of concrete and béton, the methods in use for driving piles, and the construction of cofferdams : the reader who wishes to pursue the subject further is referred to the volume of this series on "Foundations and Concrete Works," where he will find a detailed description of these operations.

RETAINING WALLS.

26. The name of *retaining wall* is applied generally to all walls built to support a mass of earth in an upright or nearly upright position ; but the term is, strictly speaking, restricted to walls built to retain an artificial bank, those erected to sustain the face of the solid ground being called *breast walls.* (See fig. 9.)

27. *Retaining Walls.*—Many rules have been given by different writers for calculating the thrust which a bank of earth exerts against a retaining wall, and for determining the form of wall which affords the greatest resistance with the least amount of material. The application of these rules to practice is, however, extremely difficult, because we have no means of ascertaining the exact manner in which earth acts against a wall ; and they are, therefore, of little

value except in determining the general principles on which the stability of these constructions depends. (See Note A, p. 155.)

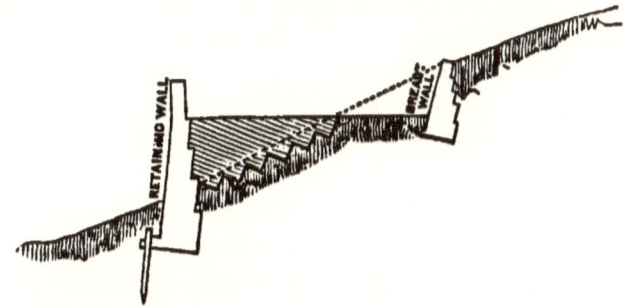

Fig. 9.

28. The calculation of the stability of a retaining wall divides itself into two parts.

1st. The thrust of the earth to be supported.

2nd. The resistance of the wall.

29. Definitions (see fig. 10).—*The line of rupture* is that along which separation takes place in case of a *slip* of earth.

Fig. 10.

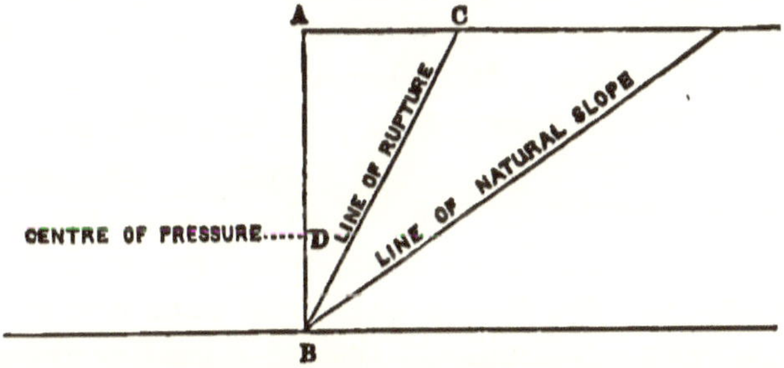

The slope which the earth would assume, if left totally un-supported, is called the *natural slope*, and it has been found that the line of rupture generally divides the angle formed by the natural slope and the back of the wall into nearly equal parts.

The *centre of pressure* is that point in the back of the wall above and below which there is an equal amount of pres-sure; and this has been found by experiment and calculation to be at $\frac{2}{3}$rds of the vertical height of the wall from its top.

The wall is assumed to be a solid mass, incapable of sliding forward, and giving way only by turning over on its front edge as a fulcrum. In the annexed diagrams the foundations of the walls have, in all cases, been omitted, to simplify the subject as much as possible. The term *slope* in the following investigation is used as synonymous with the expression *line of rupture.*

30. *Amount and Direction of the Thrust.*—There are two ways in which this may be calculated:—1st, By considering the earth as a solid mass sliding down an inclined plane, all slipping between the earth and the back of the wall being prevented by friction. This gives the *minimum* thrust of the earth   2nd, By assuming the particles of earth to have so little cohesion, that there is no friction either on the slope or against the back of the wall. This method of calculation gives the *maximum* thrust.

The real thrust of any bank will probably be somewhere between the two, depending on a variety of conditions which it is impossible to reduce to calculation; for, although we may by actual experiments with sand, gravel, and earths of different kinds, obtain data whence to calculate the thrust exerted by them in a perfectly dry state, another point must be attended to when we attempt to reduce these results to practice, viz. the action of water, which, by destroying the cohesion of the particles of earth, brings the mass of material behind the wall into a semi-fluid state, rendering its action more or less similar to that of a fluid according to the degree of saturation.

The tendency to slip will also very greatly depend on the manner in which the material is *filled* against the wall. If the ground be *benched out* (see fig. 9), and the earth well punned in layers inclined *from* the wall, the pressure will be very trifling, provided only that attention be paid to surface and back drainage. If, on the other hand, the bank be tipped in the usual manner in layers sloping *towards* the wall, the full pressure of the earth will be exerted against it, and it must be made of corresponding strength.

31. *Calculation of Minimum Thrust.*—The weight of the prism of earth represented by the triangle A B C, fig. 10,

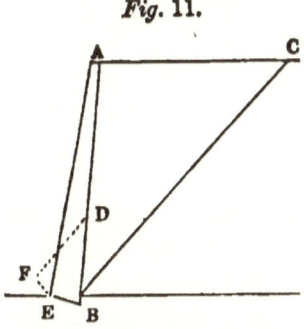

Fig. 11.

will be directly as the breadth A C, the height being constant; and the inclination of B C remaining constant, but the height varying, the weight will be as the square of the height. If, therefore, we call the weight of the prism A B C, W, the breadth A C, b, the height A B, h, and the specific gravity of the earth, s, we shall have $W = \dfrac{b\,h\,s}{2}$ If we call the thrust of W in the direction of the slope, W', then (neglecting friction), on the principle of the inclined plane, W will be to W' as the length of the incline is to its height; or, calling the length B C, l, then

$$l : h :: W : W' = \frac{h\,W}{l} = \frac{b\,h^2\,s^*}{2\,l}$$

The effect of the weight of the prism A B C to overturn the wall will be as W' multiplied by the leverage E F, fig. 11, found by letting fall the perpendicular E F, from the front edge of the wall, upon D F, drawn through the centre of pressure in a direction parallel to the slope. When D F

---

* The value of W' here given will increase with the length of A C in a constantly decreasing ratio, never exceeding $\dfrac{h^2\,s}{2}$ supposing the back of the wall to be upright. But in practice the friction must always be taken into consideration; and, as this increases directly as A C, there will be a limit at which the thrust and the resistance balance each other, this limit being the natural slope; and, as the thrust and the resistance increase with the length of A C in different ratios, there will be a point at which the effective thrust is greatest, or, in other words, a slope of maximum thrust which determines the position of the line of ruptures.

passes through E, then E F=0, and the thrust has no tendency to overturn the wall; and, when D F falls within the base of the wall, E F becomes a negative quantity, the thrust increasing its stability. Calling the overturning thrust T, we have

$$T = W' \times EF = \frac{b\,h^2\,s \times EF}{2\,l},$$

the value of E F * depending on the inclination of the slope and the width of the base of the wall

32. *Calculation of Maximum Thrust.*—If we consider the moving mass to slide freely down the slope, and the friction between the earth and the back of the wall to be so slight as to be inappreciable, then the prism A B C will act as a wedge, with a pressure perpendicular to the back of the wall, which will be the same whatever the inclination of B C, the height and inclination of the back of the wall being constant, and as the square of the height where the height varies, the pressure being the least when the back of the wall is vertical; for calling the pressure P, and drawing A I, fig 12, perpendicular to B C, we have, on the principle of the wedge,

$$AI : AB \cdot : W' : P = \frac{W' \times AB}{AI} = \frac{b\,h^2\,s \times AB}{2\,l \times AI}$$

and by construction $b\,h = l\,AI$, as they are each equal to twice the area of the triangle A B C; therefore, by substitution,

$$P = \frac{l\,AI\,h\,s \times AB}{2\,l\,AI} = \frac{h\,s \times AB}{2}$$

The effect of the prism A B C to overturn the wall will be P multiplied by the leverage E F,† which will be found by

---

$$* \quad EF = \frac{h}{l} \times \left(\frac{b}{3} - EB \right) \text{ and}$$

$$T = W' \times EF = \frac{b\,h^2\,s}{2\,l} \times \frac{h}{l}\left(\frac{b}{3} - EB \right) - \frac{b\,h^3\,s}{2\,l^2} \times \left(\frac{b}{3} - EB \right).$$

† Calling the angle X A B = θ

$$EF = \frac{AB}{3} \pm \frac{EB \cdot AX}{AB} = \frac{h}{3}\,\text{cosec.}\,\theta \pm EB\,\cos.\,\theta.$$

           And

drawing D F, fig. 13, at right angles to the back of the wall through the centre of pressure, and making E F

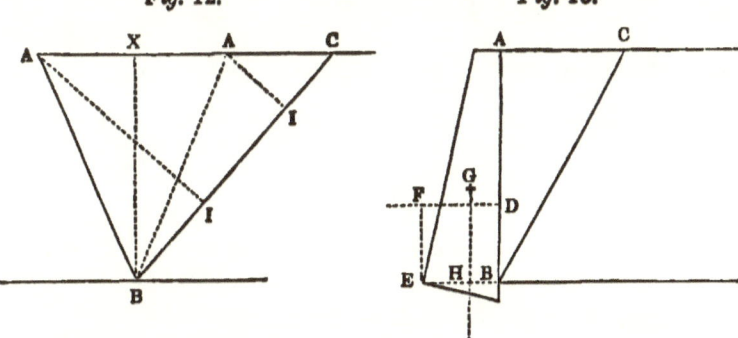

Fig. 12.   Fig. 13.

perpendicular to it; then calling the overturning thrust, as before, T,

$$T = P \times E F = \frac{A B \times h s \times E F}{2}$$

When D F passes through E, then E F=0, and the thrust has no tendency to overturn the wall; and, if D F falls within the base, the thrust will *increase* its stability. When the back of the wall is vertical, then

$$A B = h \text{ and } E F = \frac{h}{3} \text{ and } T = \frac{h^3 s}{6}.$$

33. These results show that, where the friction of the earth against the slope and the back of the wall is destroyed by the filtration of water, the action of the earth will be precisely similar to that of a column of water of the height of the wall. The pressure upon the side of any vessel is the half of the pressure that would take place upon the bottom if of the same area. Now, calling the specific gravity of the water s, the pressure upon the bottom, sup-

---

And $T = P \times E F = \frac{A B . h s}{2} \times \left(\frac{A B}{3} \pm \frac{E B . A X}{A B} \right) =$
$\frac{h s}{2} \times \left(\frac{A B^2}{3} \pm E B . A X \right)$

The positive sign is to be used when the back of the wall leans backwards; the negative, when it leans forwards.

posing its length to be A B, would be hs A B; therefore the pressure upon the side will be

$$\frac{hs\,\mathrm{A\,B}}{2}\text{; and } \mathrm{T} = \mathrm{P} \times \mathrm{E\,F} = \frac{hs\,\mathrm{A\,B.E\,F}}{2}.$$

And, where the back of the wall is vertical, then

$$\mathrm{A\,B} = h \text{ and } \mathrm{E\,F} = \frac{h}{3} \text{ as above. Therefore}$$

$$\mathrm{P} = \frac{h^2 s}{2} \text{ and } \mathrm{T} = \frac{h^2 s}{2} \times \frac{h}{3} = \frac{h^3 s}{6};$$

which results are precisely the same as those arrived at above.

34. *Resistance of the Wall.*—Considering the wall as a solid mass, the effect of its weight to resist an overturning thrust will be directly as the horizontal distance E H from its front edge to a vertical line drawn through G, the centre of gravity of the wall, fig. 13; or calling the resistance R, and the weight of the wall w, then R$=w \times$ E H. E H will be directly as E B, the proportions of the wall being constant; therefore a wall of triangular section will afford more resistance than a rectangular one of equal sectional area, the base of a triangle being twice that of a rectangle of equal height and area.

If the wall be built with a curved concave batter, fig. 14, E H will be still greater than in the case of a triangular wall of equal sectional area; and, if the wall were one solid mass incapable of fracture, this form would offer more resistance than the triangular. But, as this is not the case, we may consider any portion of the wall cut off from the bottom by a level line to be a distinct wall resting upon the lower part as a foundation.

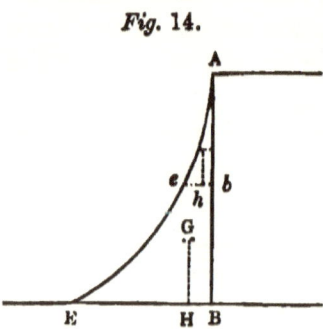

Fig. 14.

Imagine A e b to be a complete wall capable of turning upon e as a fulcrum. The resistance would be considerably less than that of the corresponding portion of a triangular wall. In the case of a triangular wall the proportions of the resistance to the thrust will be the same throughout its height. In the case of a rectangular one, the resistance will bear a greater proportion to the thrust, the greater the distance from the bottom. In the case of a wall with a concave curved batter, the reverse of this takes place.

The value of E H will be greatest when E H = E B, the wall will be then exactly balanced on H; but in practice this limit should never be reached, for fear the wall should become crippled by depending on the earth for support. The value of E H will be least when H coincides with E, which opposite limit also is never reached in practice—for obvious reasons—as the wall would in this case overhang its base, and be on the point of falling forward.

35. The increased leverage is not the only advantage gained by the triangular form of wall. In the foregoing investigation, we have considered the wall as a solid mass turning on its front edge. Now, practically, the difficulty is not so much to keep the wall from overturning as to prevent the courses from sliding on each other.

In an upright wall built in horizontal courses, the chief resistance to sliding arises from the adhesion of the mortar; but, if the wall be built with a sloping or *battering* face, the beds of the courses being inclined to the horizon, the resistance to the thrust of the bank is increased in proportion to the tendency of the courses to slide down towards the bank; thus rendering the adhesion of the mortar merely an additional security The importance of making the resistance independent of the adhesion of the mortar is obviously very great, as it would otherwise be necessary to delay backing up a wall until the mortar were thoroughly set, which might require several months.

36. The exact determination of the thrust which will be exerted against a wall of given height is not possible in practice; because the thrust depends on the cohesion of the earth. the dryness of the material, the mode of backing up the wall, and other conditions which we have no means

of ascertaining. Experience has, however, shown that the base of the wall should not be less than one-fourth, and the batter or slope not less than one-sixth of the vertical height, wherever the case is at all doubtful

37. The results of the above investigation are illustrated in figures 15, 16, 17, 18, and 19, which show the relative sectional areas of walls of different shapes, that would be required to resist the pressure of a bank of earth 12 feet high.

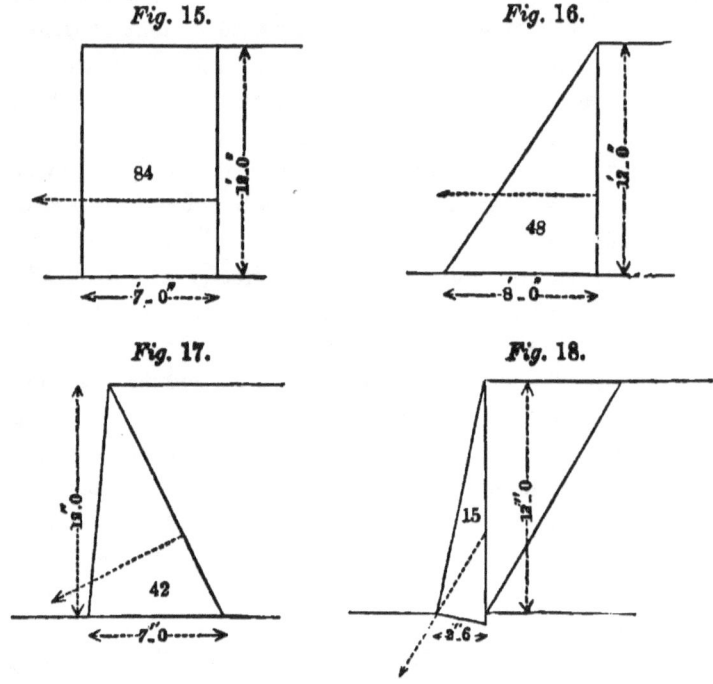

Fig. 15.   *Fig.* 16.

Fig. 17.   *Fig.* 18.

The first three examples are calculated to resist the maximum, and the fourth, the minimum, thrust; whilst the last figure (fig. 19) shows the modified form usually adopted in practice.

38. It is sometimes necessary in soft ground to protect the *toe* or front edge of a retaining wall with sheet-piling, to prevent it from being forced forward; this is shown in fig. 9.

Fig. 19.

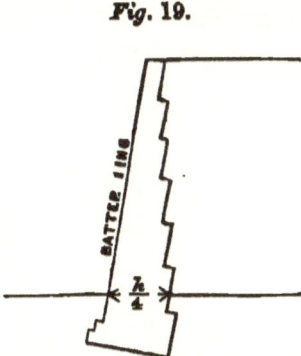

39. *Counterforts.*—Retaining walls are often built with counterforts, or buttresses, at short distances apart, which allow of the general section of the wall being made lighter than would otherwise be the case. The principle on which these counterforts are generally built is, however, very defective, as they are usually placed *behind* the wall, which frequently becomes torn from them by the pressure of the earth. The strength of any retaining wall would, however, be greatly increased were it built as a series of arches, abutting on long and thin buttresses; but the loss of space that would attend this mode of construction has effectually prevented its adoption except in a few instances.

40. *Breast Walls.*—Where the ground to be supported is firm, and the strata are horizontal, the office of a breast wall is more to protect, than to sustain the earth. It should be borne in mind that a trifling force, skilfully applied to unbroken ground, will keep in its place a mass of material which, if once allowed to move, would crush a heavy wall; and, therefore, great care should be taken not to expose the newly opened ground to the influence of air and wet for a moment longer than is requisite for sound work, and to avoid leaving the smallest space for motion between the back of the wall and the ground.

41. The strength of a breast wall must be proportionately increased when the strata to be supported incline

towards the wall, as in fig. 20 : where they incline from it,
the wall need be little more than a thin facing to protect
the ground from disintegration.

Fig. 20.

42. The preservation of the natural drainage is one of
the most important points to be attended to in the erection
of breast walls, as upon this their stability in a great measure depends.   No rule can be given for the best manner
of doing this; it must be a matter for attentive consideration in each particular case

ARCHES.

43. An arch in perfect equilibrium may be considered as
a slightly elastic curved beam, every part of which is in a
state of compression, the pressure arising from the weight
of the arch and its superincumbent load being transmitted
to the abutments on which it rests in a curved line called
the *curve of equilibrium*, passing through the thickness of the
arch.

44. The wedge-shaped stones of which a stone arch is
composed are called the *voussoirs*.   The upper surface of an
arch is called its *extrados*, and the lower surface its *intrados*
or *soffit* (see fig. 21).   Theoretically, a stone arch might give
way by the sliding of the voussoirs on each other; but in
practice the friction of the material and the adhesion of the
. mortar is sufficient to prevent this, and failure takes place

in the case of an overloaded arch by the voussoirs turning
on their edges.

Fig. 21.

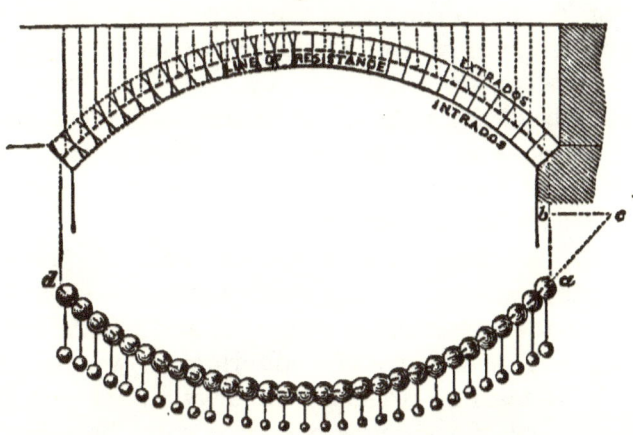

45. The curve of equilibrium will vary with the rise and
span of the arch, the depth of the arch stones, and the distribution of the load, but it will always have this property,
namely, that the horizontal thrust will be the same at every
part of it. In order that an arch may be in perfect equilibrium, its curvature should coincide with that of the curve
of equal horizontal thrust; if, from being improperly designed or unequally loaded, this latter curve approaches
either the intrados or the extrados, the voussoirs will be
liable to fracture from the pressure being thrown on a very
small bearing surface; and if it be not contained within the
thickness of the arch, failure will take place by the joints
opening, and the voussoirs turning on their edges.

46. The manner in which the curve of equilibrium is
affected by any alteration in the load placed upon an arch
may readily be seen by making an experimental equilibrated
arch with convex voussoirs, as shown in fig. 21. When
bearing its own weight only, the points of contact of the
voussoirs will lie wholly in the centre of the thickness of
the arch; when loaded at the crown, the points of contact
will approach the extrados at the crown, and the intrados

at the haunches; and, if loaded at the haunches, the reverse effect will take place.

47. If a chain be suspended at two points, and allowed
to hang freely between them, the curve it takes is the curve
of equilibrium of an arch of the same span and length on
soffit, in which the weights of the voussoirs correspond to
the weights of the links of the chain, and would be precisely the same as that marked out by the points of contact
of the curved voussoirs of an experimental arch of the same
dimensions built as above described.

48. In designing an arch, two methods of proceeding
present themselves: we may either confine the load to the
weight of the arch itself or nearly so, and suit the shape of
the arch to a given curve of equilibrium, or we may design
the arch as taste or circumstances may dictate, and load it
until the line of resistance coincides with the curve thus
determined upon.

The Gothic vaults of the middle ages were, in a great
measure, constructed on the first of these methods, being
in many cases only a few inches in thickness, and the curvature of the main ribs coinciding very nearly with their
curves of equal horizontal thrust.   We have no means of
ascertaining whether this was the result of calculation or
experiment; probably the latter, but the principle was evidently understood.

At the present day, the requirements of modern bridge
building often leave the architect little room for choice in
the proportions of his arches, or the height and inclinations
of the roadway they are to carry; and it becomes necessary
to calculate with care the proportion of the load which each
part of the arch must sustain, in order that the curve of
equilibrium may coincide with the curvature of the arch

49. The formulæ for calculating the equilibration of an arch
are of too intricate a nature to be introduced in these pages;
but the principles on which they depend are very simple

Let it be required to construct a stone arch of a given
curvature to support a level roadway, as shown in fig. 21.

and to find the weight with which each course of voussoirs
must be loaded to bring the arch into equilibrium.

Draw the centre line of the arch to a tolerably large scale
in an inverted position on a vertical plane, as a drawing
board, for instance, and from its springing points *a*, *d*, suspend a fine silk thread of the length of the centre line
strung with balls of diameter and weight corresponding to
the thickness and weight of the voussoirs of the arch ; then
from the centre of each ball suspend such a weight as will
bring the thread to the curve marked on the board, and
these weights will represent the load which must be placed
over the centre of gravity of each of the voussoirs, as shown
by the dotted lines, in order that the arch may be in equilibrium.

To find what will be the thrust at the abutments, or at
any point in the arch. draw *a c*, touching the curve, the vertical line *a b* of any convenient length, and the horizontal
line *b c*, then the lengths of the lines *a c*, *a b*, and *b c*, will
be respectively as the thrust of the arch at *a*, in the direction *a c*, and the vertical pressure and horizontal thrust into
which it is resolved; and the weight of that part of the
arch between its centre and the point *a*, which is represented by *a b*, being known, the other forces are readily calculated from it.

50. When the form of an arch does not exactly coincide
with its curve of equal horizontal thrust, there will always
be some minimum thickness necessary to contain this curve,
and to insure the stability of the arch.   In a semicircular
arch, fig. 22, whose thickness is ⅒th of its radius, the line
of equal horizontal thrust just touches the extrados at the
crown, and the intrados at the haunches, pointing out the
places where failure would take place with a less thickness
or an unequal load by the voussoirs turning on their edges.
Those arches which differ most from their curves of equal
horizontal thrust are semicircles and semi-ellipses, which
have a tendency to descend at their crowns and to rise at
their haunches, unless they are well *backed up*.   Pointed

c

arches have a tendency to *rise* at the crown; and, to prevent
this, the cross springers of the ribbed vaults of the middle

Fig. 22.

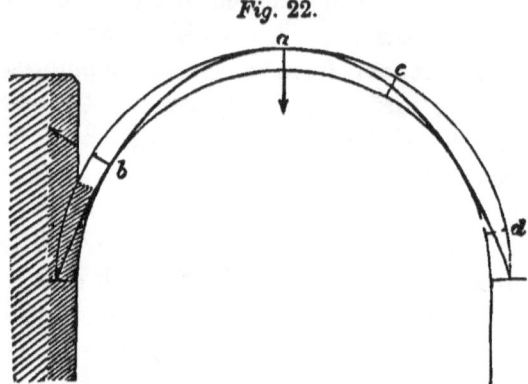

ages were often made of a semicircular profile, their flat-
ness at the crown being concealed by the bosses at their
intersections.

51. If the experiment be tried of equilibrating, in the
manner above described, a suspended semicircular or semi-
elliptical arch, it will be found to be practically impossible,
as the weight required for that purpose becomes infinite at
the springing. This difficulty does not exist in practice, for
that part of an arch which lies beyond the plane of the
face of the abutment in reality forms a part of the abutment itself (fig. 22).

The Gothic architects well understood this, and in their
vaulted roofs built this portion in horizontal courses as part
of the side walls (fig. 23), commencing the real arch at a
point considerably above the springing.

52. The depth of the voussoirs in any arch must be sufficient to contain the curve of equilibrium under the greatest
load to which it can be exposed; and, as the pressure on
the arch stones increases from the crown to the springing,
their depth should be increased in the same proportion.
Each joint of the voussoirs should be at right angles to a
tangent to the curve of equilibrium at the point through
which it passes.

Fig. 23.*

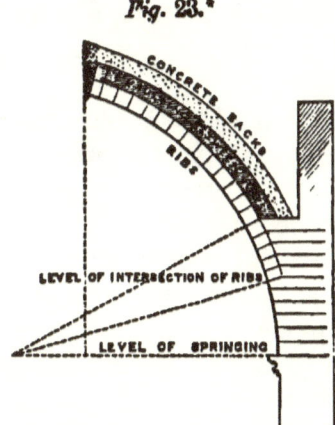

LEVEL OF INTERSECTION OF RIBS

LEVEL OF SPRINGING

53. *Brick Arches.*—In building arches with bricks of the common shape, which are of the same thickness throughout their length, a difficulty arises from the thickness of the mortar joints at the extrados being greater than at the intrados, thus causing settlement and sometimes total failure. To obviate this difficulty, it is usual to build brick arches in separate rings of the thickness of half a brick, having no connection with each other beyond the adhesion of the mortar or cement, except an occasional course of headers where the joints of two rings happen to coincide. There is, however, a strong objection to this plan, viz. that, if the curve of equal horizontal thrust do not coincide with the curvature of the arch, the line of pressure will cross the rings, and cause them to separate from each other

54. The preferable plan will be, therefore, to bond the brickwork throughout the whole thickness of the arch, using either cement or hard-setting mortar, which will render the thickness of the joints of comparatively little importance.

Cement, however, is not so well suited for this purpose as the hard-setting mortars made from the Lias limes, because it sets before the work can be completed; and in case of any settlement, however trifling, taking place on the striking of the centres, the work becomes crippled. It is therefore preferable to use some hard-setting mortar, which does not, however, set so quickly as cement, thus allowing

* This diagram is slightly altered from one of the illustrations to Professor Willis's paper "On the Construction of the Vaults of the Middle Ages," in the Transactions of the Royal Institute of British Architects, Vol. I., Part 2

the arch to adjust itself to its load, or, in technical language, to *take its bearing*, before the mortar becomes perfectly hard.

55. We have in the preceding remarks considered an
equilibrated arch as a curved beam, every part of which is
in a state of compression; and, in an arch composed of
stone voussoirs, this is practically the case.

We may, however, by the employment of other materials,
as cast iron and timber, construct arches whose forms differ
very materially from their curves of equal horizontal thrust.

Thus the semicircular arch (fig. 22), which, if built of
stone voussoirs small in proportion to the span of the arch,
would fail by the opening of the joints at *a* and *b*, might be
safely constructed with cast iron ribs, with the joints placed
at *c* and *d*, the metal at the points *a* and *b* being exposed to
a cross strain precisely similar to that of a horizontal beam
loaded in the centre.

56. Laminated arched beams, formed of planks bent
round a mould to the required curve and bolted together
have been extensively used in railway bridges of large span
during the last ten years, and from their comparative elasticity, and the resistance they offer to both tension and
compression, are very well adapted to structures of this
kind, which have to sustain very heavy loads passing with
great rapidity over them.

It is to be regretted, however, that the perishable nature
of the material does not warrant their long duration, not-
withstanding every precaution that can be taken for the
preservation of the timber.

57. *Skew Arches* —In ordinary cases the plan of an arch
is rectangular, the faces of the abutments being at right
angles to the fronts; but of late years the necessity which
has arisen on railway works for carrying communications
across each other without regard to the angle of their intersection has led to the construction of oblique or *skew* arches

58. In an ordinary rectangular arch each course is
parallel to the abutments, and the inclination of any bed
joint with the horizon will be the same at every part of it

In a skew arch it is not possible to lay the courses parallel to the abutments, for, were this done, the thrust being at right angles to the direction of the courses, a great portion of the arch on each side would have nothing to keep it from falling. In order to bring the thrust into the right direction, the courses must therefore be laid as nearly as possible at right angles to the fronts of the arch (see fig.

Fig. 24.

24), and at an angle with the abutments ; and it is this which produces the peculiarity of the skew arch. The two ends of any course will then be at different heights, and the inclination of each bed joint with the horizon will increase from the springing to the crown, causing the beds to be *winding* surfaces instead of a series of planes as in a rectangular arch. The variation in the inclination of the bed joints is called the *twist* of the beds, and leads to many difficult problems in stone-cutting, the consideration of which would be unsuited to the elementary character of this little work.

The reader who wishes to pursue the subject is referred to the volume of this series "On Masonry and S*one-cutting."

59. *Centering.*—The *centering* of an arch is the temporary framework which supports it during its erection, and is formed of a number of ribs or *centres*, on which are placed the planks or *laggings* on which the work is built.

60. In designing centres, there are three essential points to be kept in view 1st, that there should be sufficient strength to prevent any settlement or change of form during the erection of the arch. 2nd, that means should be provided for *easing* or lowering the centre gradually from under any part of the arch. 3rd, that, as the construction of centres generally involves the use of a large quantity of timber merely for a temporary purpose, all unnecessary injury to it should be avoided, in order that its value for subsequent use may be as little diminished as possible.

61. Fig. 25 represents the construction of the centering

used in the erection of the Gloucester Over Bridge, designed by Mr. Cargill, the contractor, which fulfils the
above-named conditions in a very perfect manner : by means
of the *striking wedges* under the radiating struts, any part of
these centres can be lowered at pleasure, and, from the
position of the struts, there is no tendency to alteration in
the curve from undue pressure on the haunches during the
erection of the arch. (See Note B, p. 155.)

62. Centering on the same principle as the above was
made use of in the erection of the Grosvenor Bridge at
Chester, by Mr. Trubshaw, the contractor for that work
Instead of the centres being made to rest on the striking

Fig. 25.

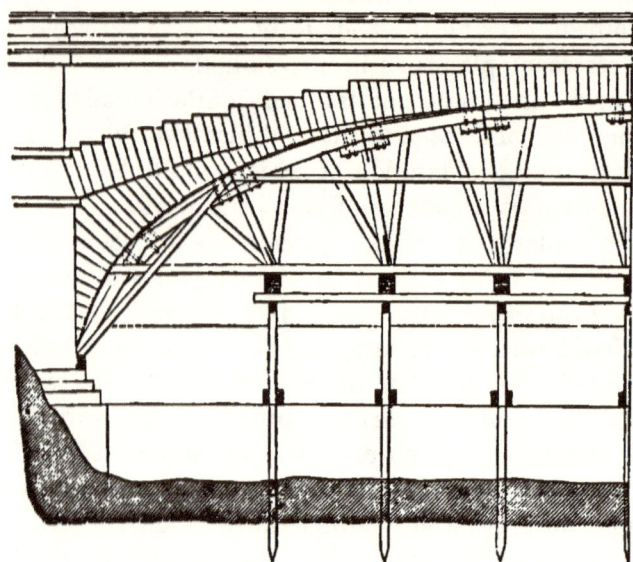

wedges, however, as in the centering for the Gloucester
Bridge, the wedges were placed under the laggings themselves, by which means the arch could be eased in the most
gradual manner

63. Where the circumstances of the case do not admit
of piles or other supports being placed between the piers, it
becomes necessary to construct a trussed framing resting

on the piers, and of sufficient strength to support the weight of the arch. The tendency of this form of centre to rise at the crown, from the great pressure thrown upon the haunches during the erection of the arch, renders it necessary to weight the crowns with blocks of stone until it is nearly completed. Centres of this kind are always costly, afford less facilities for easing, and are in every way inferior to those we have described as used at Gloucester and Chester.

64. *Abutments.*—The tendency of any arch to overturn its abutments, or to destroy them by causing the courses to

Fig. 26.

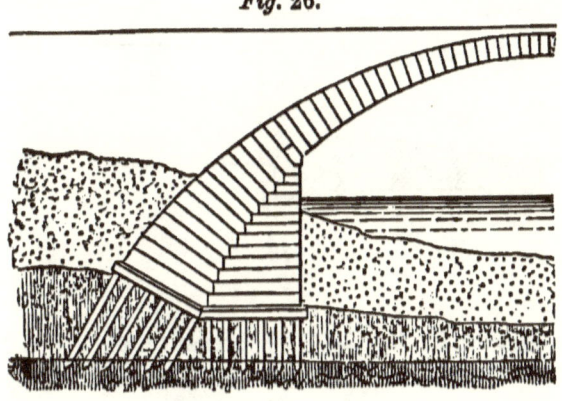

slide over each other, may be counteracted in three ways 1st, the arch may be continued through the abutment until it rests on a solid foundation, as in fig. 26 2nd, by build-

Fig. 27.

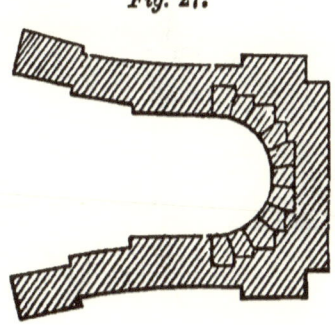

ing the abutments so as to form a horizontal arch, the thrust being thrown on the wing walls, which act as buttresses (fig. 27). 3rd, where neither of these expedients is practicable, by joggling the courses together with bed-dowel joggles, so as to render the whole abutment one solid mass (fig. 36).

65 *Wing Walls.*—Where the wing walls of a bridge are

Fig. 28.

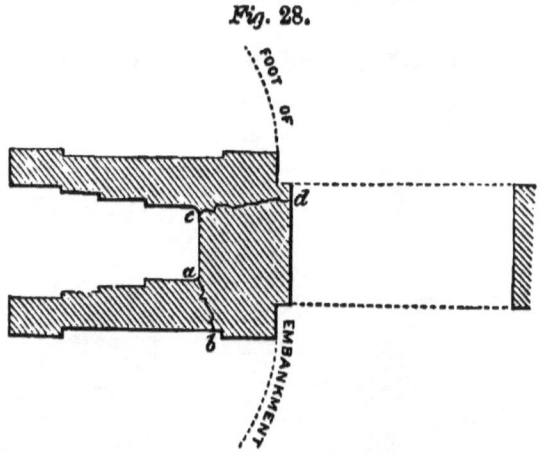

built as shown in fig. 28, the pressure of the earth will always have a tendency to fracture them at their junction with the abutments, as shown by the lines *a b, c d.* Equal

Fig. 29.

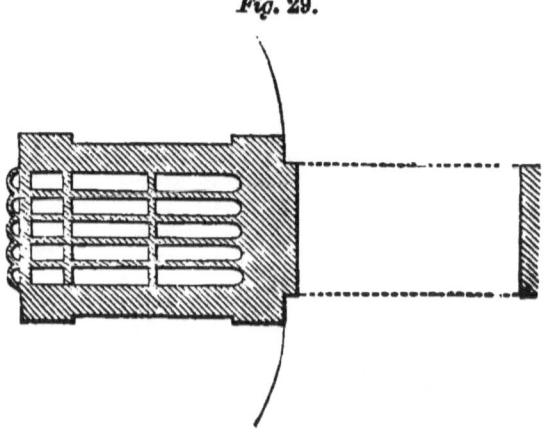

strength with the same amount of material will be obtained by building a number of thin longitudinal and cross walls, as shown in fig. 29, by which means, the earth being kept from the back of the walls, there is no tendency to failure of this kind.

66. *Vaulting.*—The ordinary forms of vaults may be classed under three heads, viz. *cylindrical, coved,* and *groined.*

A *cylindrical* vault is simply a semicircular arch, the ends of which are closed by upright walls, as shown in fig. 30. When a vault springs from all the sides of its plan, as in fig. 31, it is said to be *coved.* When two cylindrical vaults intersect each other, as in fig. 32, the intersections of the

Fig. 30. *Fig.* 31.

vaulting surfaces are called *groins,* and the vault is said to be *groined.*

67. In the Roman style of architecture, and in all common vaulting, the vaulting surfaces of the several compartments are portions of a continuous cylindrical surface, and the profile of a groin is simply an oblique section of a semi-cylinder.

68. Gothic ribbed vaulting is, however, constructed on a totally different principle. It consists of a framework of light stone ribs supporting thin pannels, whence this mode of construction has obtained the name of *rib and pannel* vaulting. The curvature of the diagonal ribs or cross springers, and of the intermediate ribs, is not governed in any way by the form of the transverse section of the vault, and in this consists the peculiarity of ribbed vaulting. This will be understood by a comparison of figs. 32 and 33. For a description of the several varieties of Gothic vaults, and the modes of tracing the curves of the ribs, the reader is referred to the volume of this series on " Masonry and Stone-cutting."

c 3

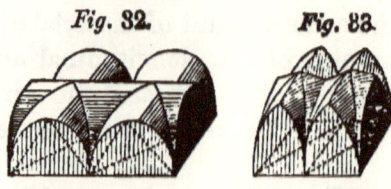

Fig. 32.   *Fig. 33.*

Roman vaulting.   Gothic vaulting.

69. Domes are vaults on a circular plan. The equilibrium of a dome depends on the same conditions as that of a common arch, but with this difference, that, although a dome may give way by the weight of the crown forcing out the haunches, failure by the weight of the haunches squeezing up the crown is impossible, on account of the support the voussoirs of each course receive from each other.

MASONRY—BRICKWORK—BOND

70. The term *masonry* is sometimes applied generally to all cemented constructions, whether built of brick or stone ; but in England the use of the term is confined exclusively to stone-work.   .

71. There are many kinds of masonry, each of which is known by some technical term expressive of the manner in which the stone is worked; but they may all be divided under three heads.

1st. Rubble work (fig. 34), in which the stones are used without being squared

2nd. Coursed work (fig. 35), in which the stones are squared, more or less, sorted into sizes, and ranged in courses.

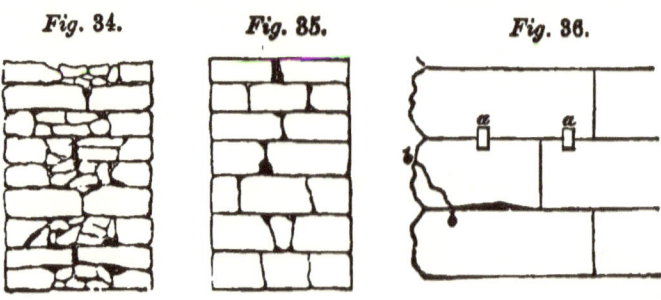

Fig. 34.   *Fig. 35.*   *Fig. 36.*

3rd. Ashlar work* (fig. 36), in which each stone is squared and dressed to given dimensions.

72. Different kinds of masonry are often united. Thus a wall may be built with ashlar facing and rubble backing; and there are many gradations from one class of masonry to another, as *coursed rubble*, which is an intermediate step between rubble work and coursed work.

73. In ashlar masonry, the stability of the work is independent, in ordinary cases, of the adhesion of the mortar. Rubble work, on the contrary, depends for support in a great measure upon it.

74. In dressing the beds of ashlar work, care must be taken not to work them hollow, so as to throw the pressure upon the edges of the stones, as this leads to unsightly fractures, as *b b*, fig. 36.

75. Where there is a tendency of the courses to slide on each other from any lateral pressure, it may be prevented by bed-dowel joggles, as shown at *a a*, fig. 36

76 Where the facing and the backing of a wall do not contain the same number of courses, as in the case of a brick wall with stone facings (fig. 37), the work will be liable to settle on the inside, as shown by the dotted lines, from the greater number of mortar joints. The only way of preventing this is to set the backing in cement, or some hard and quick-setting mortar

Fig. 37.

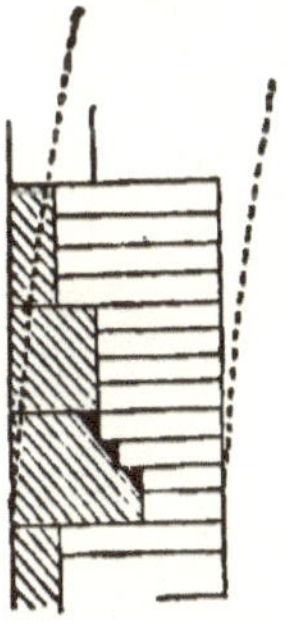

77. In facing brickwork with stone ashlar, the stones should be all truly squared, and worked to sizes that will bond with the brickwork. If this be neglected, there will be numerous vacuities in the thickness of the wall (see fig. 37) and the facing and backing will have a tendency to separate

78. *Bond*, in masonry, consists in the placing of the stones in such relative positions that no joint in any course shall be in the same plane with any

* In London the term "ashlar" is commonly applied to a thin facing of stone placed in front of brickwork.

other joint in the course immediately above or below it
This is called *breaking joint*

79. Stones placed lengthwise in any work are called
stretchers, and those placed in a contrary direction are called
headers. When a header extends through-
out the whole thickness of a wall, it is
called a *through*.

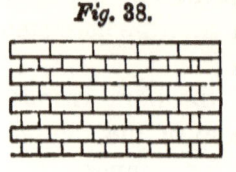

Fig. 38.

Fig. 39.

79. There are two kinds of bond made
use of by bricklayers, called respectively
English bond and *Flemish bond*. In the
first the courses are laid alternately with
headers and stretchers (fig. 38); in the
second, the headers and stretchers alter-
nate in the same course (fig. 39). This
is considered to have the neatest appear-
ance: but, as the number of headers required is fewer than
in English bond, there is not so much lateral tie, and on
this account it is considered to be much inferior to it in
strength. A common practice, which cannot be too much
reprobated, is that of building brick walls with two qualities
of bricks, without any bond between them, the headers of
the facing bricks being cut in two to save the better material, thus leaving an upright joint between the facing and
backing.

80. In building upright walls which have to sustain a
vertical pressure, three leading principles must be kept in
view.

1. Uniformity of construction throughout the whole
thickness.

2. The bonding of the work together

3. The proper distribution of the load.

81 *Uniformity of Construction.*—We have already spoken
of the danger arising from the backing of a wall containing
more compressible material than the facing; but it cannot
be too often repeated, that in all building operations it is
not the *amount*, but the *irregularity* of settlement which is
so dangerous. Thus a rubble wall, with proper care, may

be carried up to a great height, and bear safely the weight of the floors and roof of a large building, whilst a wall built of bricks and mortar, and faced with dressed ashlar, will, under similar circumstances, be fractured from top to bottom, from the difference in settlement of the facing and backing

It is a common but vicious practice to build the ends of joists and other timbers into the walls, and to rest the superincumbent work upon them. This is liable to lead to settlements from the shrinking of the timber, and should always be guarded against by leaving proper recesses for the ends of the timbers, so that the strength of the masonry or brickwork shall be quite independent of any support from them.

82. *Bond.*—In addition to the bonding together of the materials above described, a further security against irregular settlement is usually provided for brick walls, in the shape of ties of timber, called *bond*, which are cut of the depth and thickness of a brick, and built into the work. There is, however, a great objection to the use of timber in the construction of a wall, as it shrinks away from the rest of the work, and often endangers its stability by rotting.

83. Instead of bond timbers, hoop-iron bond is now very generally used. This is formed of iron hooping, tarred, to protect the iron from contact with the mortar, and laid in the thickness of the mortar joints. This forms a very perfect longitudinal tie, and has all the advantages, with none of the disadvantages, of bond timbers.

84. *Distribution of the Load.*—It is always advisable, when a heavy load has to be supported on a few points, as in the case of a large floor resting on girders, to bring the weight as nearly as possible on the centre of the wall, and to distribute it over a large bearing surface, by stone bonding through its whole thickness; this arrangement is shown in figures 40 and 41.

85. It is of importance in designing buildings to arrange the apertures for doors, windows, &c., in the different floors,

so that openings shall be over openings, and piers over piers; if this be not attended to, it is scarcely possible to

Fig. 40.

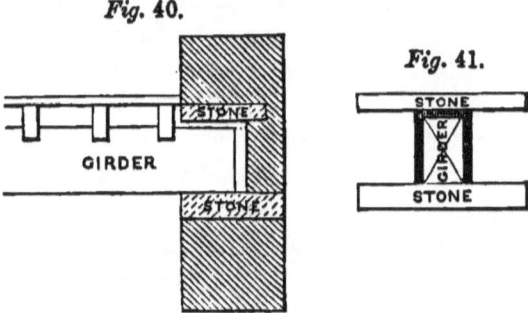

Fig. 41.

prevent settlements   In addition to this, as the pressure on the foundations will be greatest under the piers, it is desirable to connect these with inverted arches, by which means the weight is distributed equally over the whole surface of the foundations.

86. All openings in walls for doors, windows, gateways, &c., should be arched over throughout the whole thickness of the walls in which they occur; and wooden lintels and bressummers should only be introduced as ties to counteract the thrust of the arches, and as attachments for the internal finishings.

87. Bressummers of cast iron are often used for supporting the walls of houses over large openings, as in the case of shop fronts; but they have the disadvantage of being liable to be cracked, in case of fire, if water is thrown on them whilst in a heated state, which renders their use very objectionable, as no dependence can be placed upon them after having been suddenly cooled in this manner, even if they do not actually break at the time.

PARTITIONS.

88. The partitions forming the interior divisions of a building may be either solid walling of brick or stone, or they may be constructed entirely of timber, or they may be frames of timber filled in with masonry or brickwork.

It will always be best, both for durability and security against fire, to make the partitions of solid walling; but this is not always practicable, and, in the erection of dwelling houses, they are for the most part made of timber.

The principles to be kept in view in the construction of framed timber partitions are very simple. Care must be taken to avoid any settlement from cross strain, and they should not in any way depend for support upon subordinate parts of the construction, but should form a portion of the

Fig. 42.

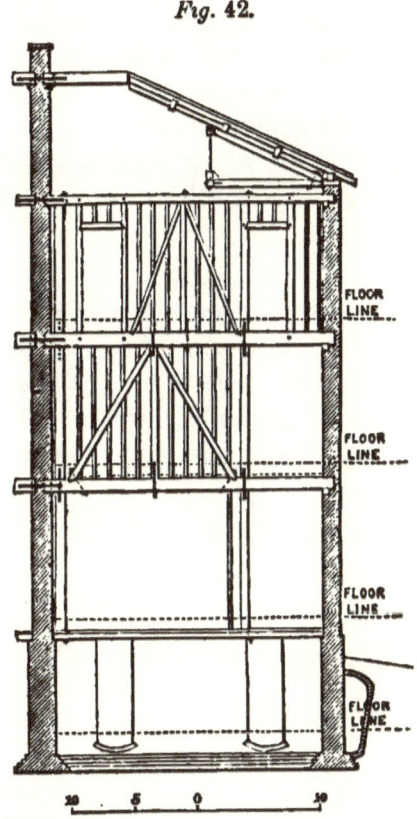

main carcase of the building, and be quite independent of the floors, which should not support, but should be supported by them.

Where a partition extends through two or more stories of a building, it should be as much as possible a continuous piece of framing, with strong sills at proper heights to support the floor joists.

Where openings occur, as for folding doors, or where a partition rests on the ends of the sill only, it should be strongly trussed, so that it is as incapable of settlement as the walls themselves. From want of attention to these points, we frequently see in dwelling-houses floors which have sunk into curved lines, doors out of square, cracked ceilings and broken cornices, and gutters that only serve to conduct the roof water to the interior of the building, to the injury of ceilings and walls, and the great discomfort of its inmates. The above remarks will be better understood by a study of fig. 42, which is an example of a framed partition extending through three stories of a dwelling house.

FLOORS.

89. The assemblage of timbers forming any *naked flooring* may be either *single* or *double*. Single flooring is formed with joists reaching from wall to wall, where they rest on *plates* of timber built into the brickwork, as in fig. 43. The floor boards are nailed over the upper edges of the joists,

Fig. 43.

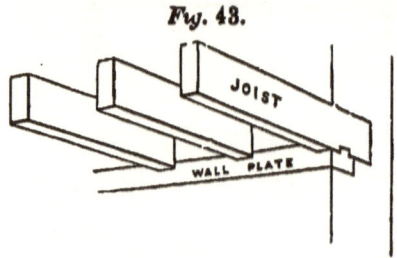

Single flooring.

whose lower edges receive the lathing and plastering of the ceilings. Double floors are constructed with stout *binding joists*, a few feet apart, reaching from wall to wall, and supporting *ceiling joists* which carry the ceiling; and *bridging joists*, on which are nailed the floor boards (fig. 44).

Iu *double-framed flooring*, the binders, instead of resting in the walls, are supported on *girders*, as shown in fig. 45

Fig 44.

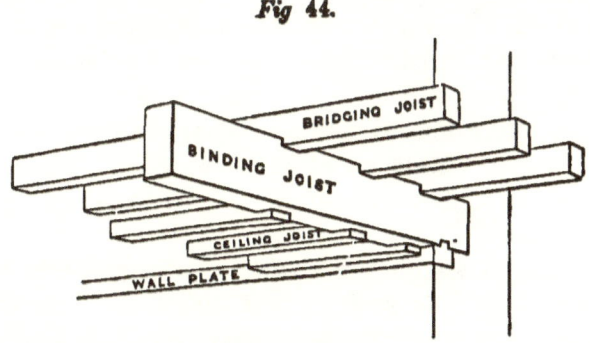

Double flooring.

Single flooring is, in many respects, inferior to double flooring, being liable to *sag*, or deflect, so as to make the

Fig. 45.

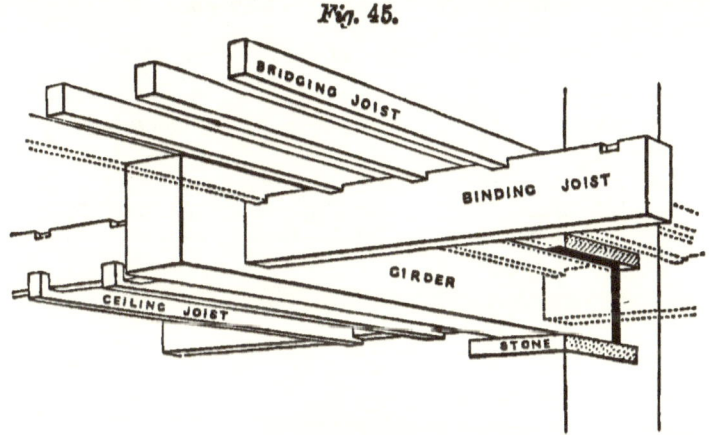

Double-framed flooring.

floor concave, and the vibration of the joists occasions injury to the ceilings, and also shakes the walls. In double flooring the stiffness of the binders and girders prevents both deflection and vibration, and the floors and ceilings *hold their lines*, that is, retain their intended form much better than in single flooring.

90 The joists in a single floor are usually laid on a plate

built into the wall, as shown in fig. 43; it is, however, preferable to rest the plate on projecting corbels, which prevents the wall being crippled in any way, by the insertion of the joists. The plates of basement floors are best supported on small piers carried up from the footings. This is an important point to be attended to, as the introduction of timber into a wall is nowhere likely to be productive of such injurious effects as at the foundations, where, from damp and imperfect ventilation, all wood-work is liable to speedy decay.

The ends of all girders should rest in recesses, formed as shown in figs. 40 and 41, and with a space for the free circulation of air round the timber, which is one of the best preventives of decay.

The manner in which ceiling joists and bridging joists are framed to the binders, and these latter tenoned into the girders, is shown in figs. 46, 47, 48, and 49.   .

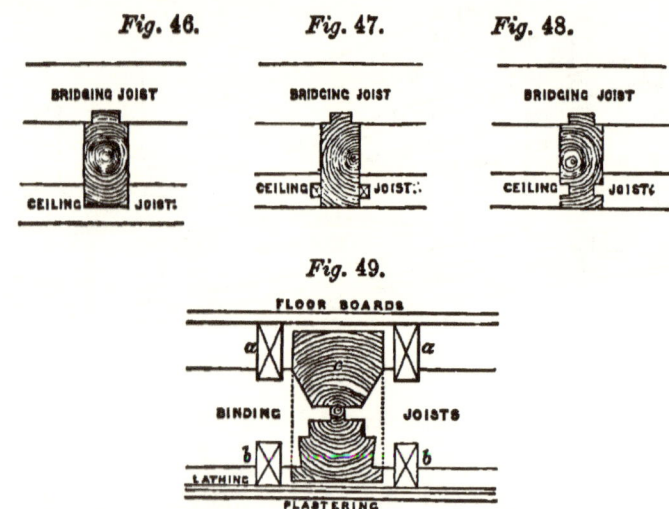

a a, bridging joists; *b b*, ceiling joists; *c*, girder.

91. Fire-proof floors are usually constructed with iron girders a short distance apart, which serve as abutments for a series of brick arches, on which either a wooden or plaster floor may be laid (see fig. 50). (See Note C, p. 156.)

Fig. 50.

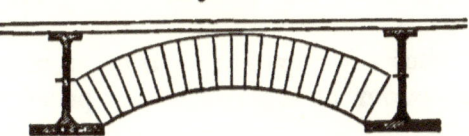

92. Of late years many terraces and flat roofs have been constructed with two or more courses of plain tiles, set in cement, and breaking joint with each other, supported at short intervals by cast-iron bearers, as shown in fig. 51. This mode of construction, although appearing very slight,

Fig. 51.

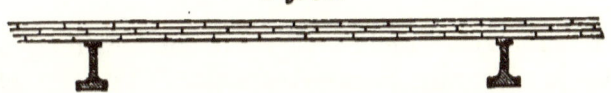

possesses great strength, and is now very much used in and about London

ROOFING.

93. In roofs of the ordinary construction, the roof covering is laid upon *rafters* supported by horizontal *purlins*, which rest on upright *trusses* or frames of timber, placed on the walls at regular distances from each other. Upon the framing of the trusses depends the stability of the roof, the arrangement of the rafters and purlins being subordinate matters of detail. The timbering of a roof may be compared to that of a double-framed floor, the trusses of the former corresponding to the girders of the latter, the purlins to the binders, and the rafters to the joists.

Timber roofs may be divided under two heads—

1st. Those which exert merely a vertical pressure on the walls on which they rest.

2nd. Those in which advantage is taken of the strength of the walls to resist a side thrust, as in many of the Gothic open timbered roofs.

94. *Trussed Roofs, exerting no Side Thrust on the Walls.*— In roofs of this kind each truss consists essentially of a pair of principal rafters or *principals*, and a horizontal *tie*

beam and in large roofs these are connected and strength
ened by *king and queen posts* and *struts* (see figs. 53 and
54).

Fig. 52 shows a very simple truss in which the tie is
above the bottom of the feet of the principals, which is

Fig. 52.

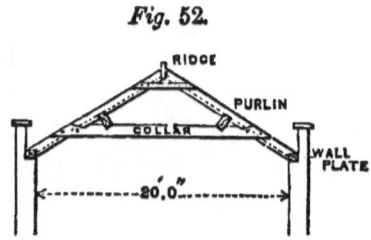

often done in small roofs for the sake of obtaining height
The tie in this case is called a *collar*. The feet of both
common and principal rafters rest on a *wall plate*. The
purlins rest on the collar, and the common rafters but
against a *ridge* running along the top of the roof. This
kind of truss is only suited to very small spans, as there is
a cross strain on that part of the principals below the collar
which is rendered harmless in a small span by the extra
strength of the principals, but which in a large one would
be very likely to thrust out the walls.

95 In roofs of larger span the tie beam is placed below
the feet of the principals, which are tenoned into, and
bolted to it. To keep the beam from *sagging*, or bending
by its own weight, it is suspended from the head of the

Fig. 53.

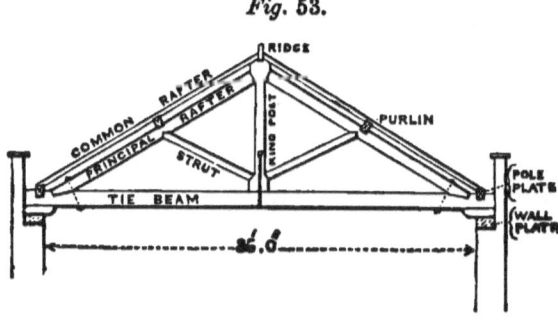

principals by a king post of wood or iron. The lower part of the king post affords abutments for struts supporting the principals immediately under the purlins, so that no cross strain is exerted on any of the timbers in the truss, but they all act in the direction of their length, the principals and struts being subjected to compression, and the king post and tie beam to tension. Fig. 53 shows a sketch of a king truss. The common rafters but on a *pole plate*, the tie beams resting either on a continuous plate, or on short templates of wood or stone.

96. Where the span is considerable, the tie beam is supported at additional points by suspension pieces called queen posts (fig. 54), from the bottom of which spring

Fig. 54.

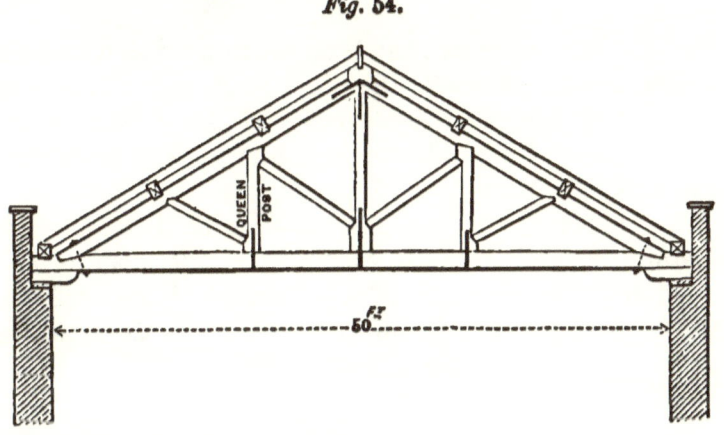

additional struts ; and, by extending this principle *ad infinitum*, we might construct a roof of any span, were it not that a practical limit is imposed by the nature of the materials. Sometimes roofs are constructed without king posts, the queen posts being kept apart by a straining piece. This construction is shown in fig. 55, which shows the design of the old roof (now destroyed) of the church of St. Paul, outside the walls, at Rome. This truss is interesting from its early date, having been erected about 400 years ago ; the trusses are in pairs, a king post being keyed

Fig. 55.

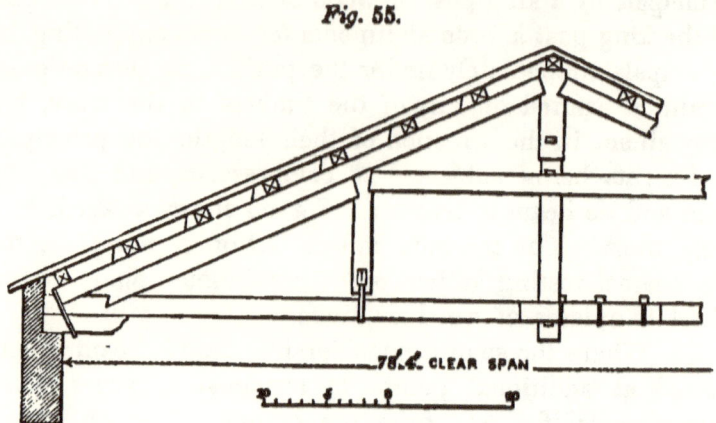

in between each pair to support the tie beams in the centre.

97. Of late years iron has been much used as a material for the trusses of roofs, the tie beams and suspending pieces being formed of light rods, and the principals and struts of rolled T or angle iron, to which sockets are riveted to receive the purlins.

The iron roofs of the new Houses of Parliament at Westminster are admirable examples of this mode of construction. The principle of the trussing of the roof over the House of Peers is shown in fig. 56   The tie beam and

Fig. 56.

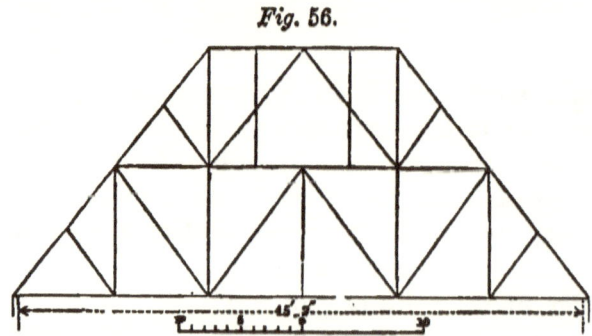

suspension rods are of flat bar iron, the principal and common rafters are of rolled T iron, the struts and purlins are of cast iron, and the whole is fitted together with cast-iron shoes.

98. The great novelty in the construction of the roofs just mentioned consists in their covering, which is formed of galvanized sheets of cast iron, lapping over each other at the joints, and forming a very perfect and water-tight covering, which is at the same time perfectly fire-proof,

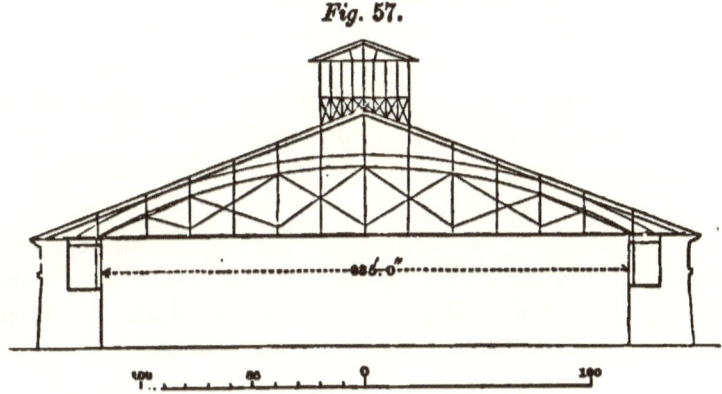

Fig. 57.

and not liable to be affected by exposure to the atmosphere.

99. The largest roof ever executed in one span is that of the Imperial Riding-House at Moscow, built in 1790, of which the span is 235 ft. (fig. 57). The principal feature in this roof is an arched beam, the ends of which are kept from spreading by a tie beam, the two being firmly connected by suspension pieces and diagonal braces: the arched beam (fig. 58) is formed of three thicknesses of

Fig. 58.

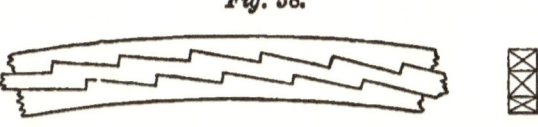

timber, notched out to prevent their sliding on each other, — a method which is objectionable on account of the danger of the splitting of the timber under a considerable strain. (See Note D, p. 156.)

100. The principle of the *bow suspension truss*, as this system of trussing is called, has been much used within the last ten years for railway bridges and similar works. One of the best executed works of this kind is a bridge over the River Ouse, near Downham Market, in Norfolk, on the line of the Lynn and Ely Railway, the trusses of which are 120 ft. span.

101. *Roofs on the principle of the Arch.* — In the 16th century, Philibert de Lorme, a celebrated French architect, published a work in which he proposed to construct roofs and domes with a series of arched timber ribs in place of trusses, these ribs being formed of planks in short lengths, placed edgewise, and bolted together in thicknesses, breaking joint (fig. 59). This mode of construction has been more or less used ever since the time of its author. An instance of its successful application on a large scale was the original dome of the Halle au Blé, at Paris, 120 ft. in diameter, built by Messrs. Legrand and Molino. This roof

Fig. 59.

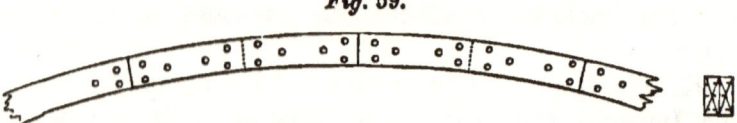

has since been replaced by an iron one, the original dome having been destroyed by fire.

The roof of the central compartment of the Pantheon Bazaar in Oxford Street, London, 38 ft. span, is another very elegant example.

102. There are, however, some great disadvantages connected with this system. There is considerable waste of material; the labour is great as compared with roofs of similar span of the ordinary construction; and, as the chief strength of the rib depends upon the lateral cohesion of the fibres of the wood, it is necessary to provide such an amount of surplus strength as shall insure it against the greatest cross strain to which it can be exposed from violent winds or otherwise

108. Struck by these disadvantages, Colonel Emy, a French military engineer, proposed, in 1817, an improvement on the system of Philibert de Lorme, which was precisely the laminated arched rib so much in use at the present day. It was not until 1825 that he obtained permission to put his design into execution in the erection of a large roof 65 ft. span at Marac, near Bayonne (fig. 60).

Fig. 60.

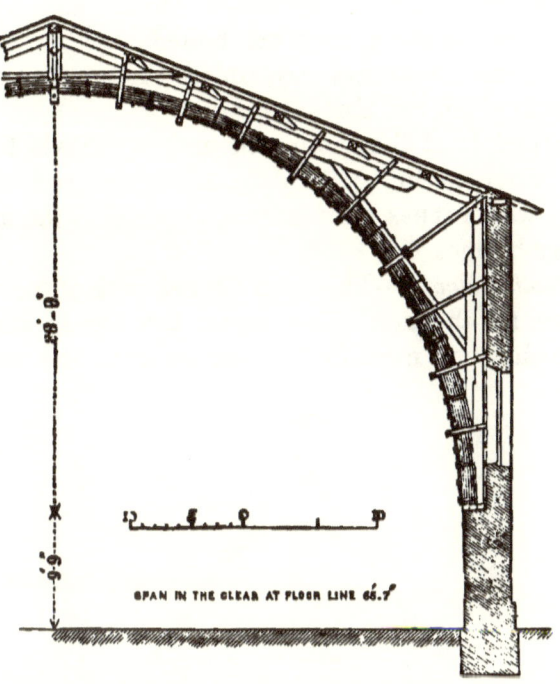

SPAN IN THE CLEAR AT FLOOR LINE 65.7'

The ribs in this roof are formed of planks bent round on templets to the proper curve, and kept from separating by iron straps, and also by the radiating struts which are in pairs, notched out so as to clip the rib between them.

The principle of the roof is exceedingly good. The principals, wall-posts, and arched rib, form two triangles, firmly braced together, and exerting no *thrust* on the walls; and the weight of the whole roof being thrown on the walls at the feet of the ribs, and not at the pole plate, the

D

walls are not tried by the action of a heavy roof, and the consequent saving in masonry is very great.*

The great difference in principle between the arched rib of Philibert de Lorme, and the laminated rib of Colonel Emy, is, that in the latter the direction of the fibre of the wood coincides with the curvature of the rib; and, as a consequence of this, the joints are much fewer; the rib possesses considerable elasticity, so as slightly to yield rather than break under any violent strain; and, from the manner in which the planks are bolted together, it is impossible for the rib to give way, unless the force applied be sufficient to crush the fibres.

The principle of the laminated arched rib was first applied in England in 1837 by the Messrs. Green of Newcastle, by whom it has been extensively used in the erection of railway bridges.

104. *Gothic Roofs.*—The open timber roofs of the middle ages come, for the most part, under the second class, viz. those which exert more or less thrust upon the walls, al-

Fig. 61.

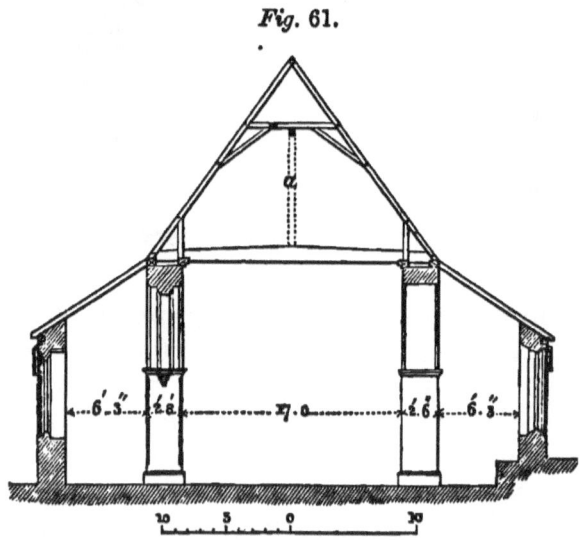

* See Tredgold's Carpentry, new and much improved edition, in 4to, 1853.

though there are many fine examples in which this is not the case.

We propose to describe the principal varieties of these roofs, without reference either to their decorative details, or to their chronological arrangement, our object here being simply to explain the principles on which they were constructed.

105. Fig. 61, which is a section of the parish church of Chaldon, near Merstham in Surrey, shows a system of roofing formerly very common. This may be compared to single flooring, as there are no principals, purlins, or even ridge. It is a defective form of roof, as the rafters have a tendency to spread and thrust out the walls. In the example before us, this effect has been prevented by the insertion of tie beams, from which the collars have been propped up (fig. 62), thus, in fact, balancing the roof on the centres of the collars, which are in consequence violently strained.

Fig. 62.

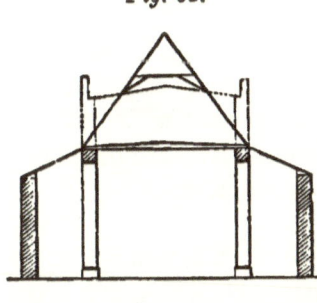

a post; b sill; c c struts.

106. After the introduction of the 4-centred arch, a great many church roofs of the construction just described were altered, as shown by the dotted lines in fig. 63, in order to obtain more light by the introduction of clerestory windows over the nave arches. The flat roofs which superseded the former ones were often formed without any truss whatever, being simply an arrangement of main beams, purlins, and rafters, precisely similar to a double-framed floor, with the difference only that the main beams, instead of being perfectly straight, were usually cut out of crooked timber so as to divide the roof into two inclined planes.

Fig. 63.

To throw the weight of the roof as low down as possible,

the ends of the main beams are often supported on upright

Fig. 64.

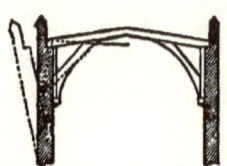

posts placed against the walls and resting on projecting corbels, the wall posts
and beams being connected by struts
in such a way that deflection in the
centre of the beam cannot take place,
unless the load be sufficient to force
out the walls, as shown by the dotted
lines in fig. 64. The struts are often cut out of stout plank,
forming solid spandrils, the edges of which are moulded to
suit the profile of the main beam (see fig. 65), which also

Fig. 65.

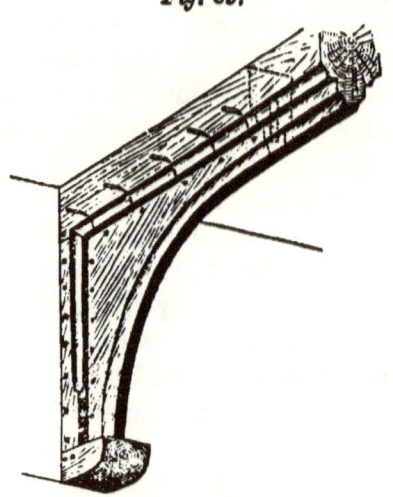

shows the manner of securing the struts to the wall posts
and to the beam with *tongues* and wooden pins. A very
good example of this construction is shown in fig. 66,
which is from West Bridgeford Church, Nottinghamshire.
There are many very beautiful examples remaining in different parts of the country.

107. A somewhat similar construction to that last described is shown in fig. 67, in which principals are introduced, strutted up from the main beam, so as to give a
greater slope to the roof than could well be obtained with a
single beam.

Fig. 66.

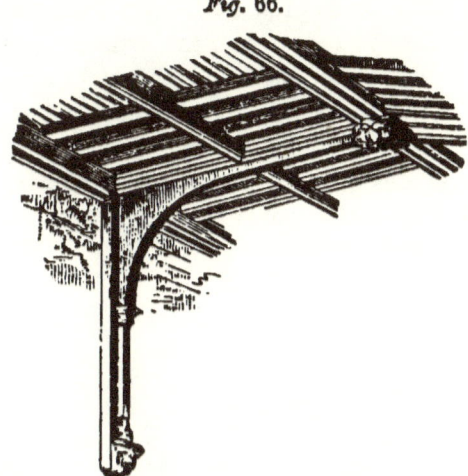

Fig. 67.

Fig. 68.

108. Fig. 68 exhibits a construction often to be met with, which, in general appearance, resembles a trussed king post roof, but which is in reality very different, the tie beam being a strong girder supporting the king post, which, instead of serving to suspend the tie beam from the principals, is a prop to the latter. In this and the previous example, any tending to deflection of the tie beam is prevented by struts: the weight of the roof is thrown by means of wall posts considerably below the feet of the rafters, so that the weight of the upper part of the wall is made available to resist the thrust of the struts.

109 The roofs we have been describing are not to be recommended as displaying any great amount of constructive skill. Indeed, although they answer very well for small spans with timbers of large scantling and side walls of sufficient thickness to resist a considerable thrust, they are totally unsuited to large spans, and are in every way inferior to trussed roofs.

The above remarks do not apply to the high pitched

roofs of the large halls of the fifteenth and sixteenth cen-
turies, which, for the most part, are trussed in a very perfect manner, so as to exert no thrust upon the walls;
although, in some instances, as at Westminster Hall, they
depend upon the latter for support.

The general design of these roofs is shown in figs. 69
and 70. The essential parts of each truss are, a pair of

Fig. 69.

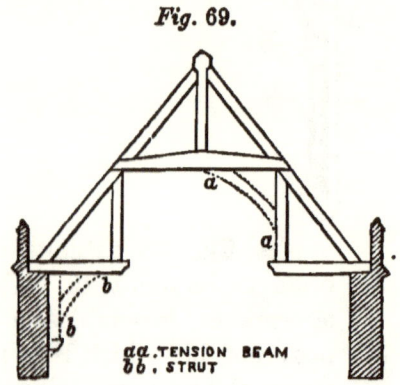

principals connected by a collar or *wind beam*, and two
hammer beams, with queen posts over them, the whole
forming three triangles, which, if not secured in their
relative positions, otherwise than by the mere transverse
strength of the principals, would turn on the points *c c* (fig.
70), the weight of the roof thrusting out the walls in the
manner shown in the figure. There are two ways in which

Fig. 70.

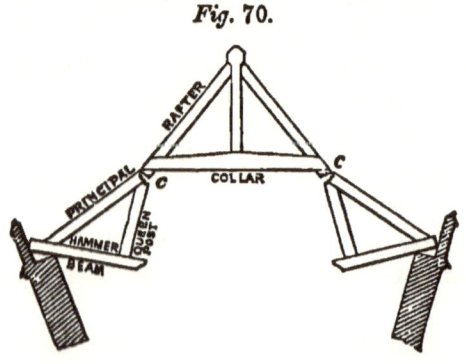

a truss of this kind may be prevented from spreading. 1st. The ends of the hammer beams may be connected with the collar by tension pieces, *a a* (fig. 69), by which the thrust on the walls will be converted into a vertical pressure. 2nd. The hammer beams may be kept in their places by struts, *b b*, the walls being made sufficiently strong by buttresses, or otherwise, to resist the thrust.

In existing examples, we find sometimes one and sometimes the other of these plans followed; and occasionally both methods are combined in such a manner that it is often difficult to say what parts are in a state of compression, and what are in a state of tension.

110. The roof of the great hall at Hampton Court (fig. 71) is very strong, and so securely tied, that were the bottom

Fig. 71.

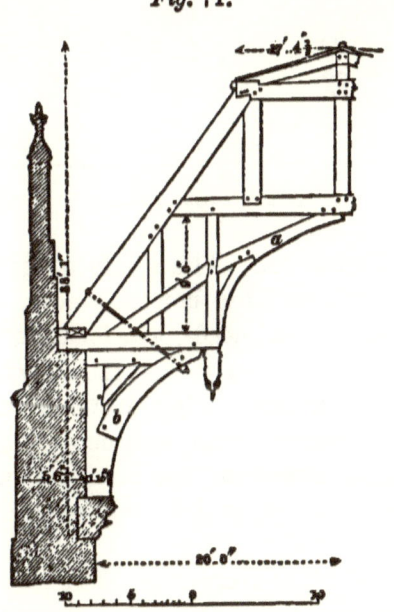

struts, *b b*, removed, there would be little danger of the principals thrusting out the walls; and, on the other hand, from the weight of the roof being carried down to a con-siderable distance below the hammer beams by the wall posts, the walls themselves offer so much resistance to side thrust, that there would be no injurious strain on them wore the tension pieces, *a a*, removed.

111. The construction of the roof of the hall at Eltham Palace, Kent (fig. 72), differs very considerably from that of

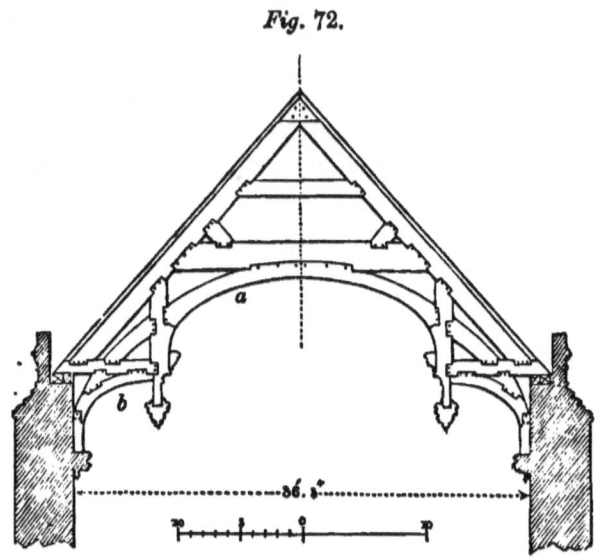

Fig. 72.

the Hampton Court roof. The whole weight is thrown on the top of the wall, and the bottom pieces, *b b*, are merely ornamental, the tension pieces, *a a*, forming a complete tie. This has been shown by a partial failure which has taken place. The wall plates having become rotten in consequence of the gutters being stripped of their lead, the weight has been thrown on the pseudo struts, which have bent under the pressure, and forced out the upper portion of the walls.

112. The roof of Westminster Hall (fig. 73) is one of the finest examples now existing of open timbered roofs. The peculiar feature of this roof is an arched rib in three thicknesses, something on the principle of Philibert de Lorme ; but it is so slight, compared with the great span, that it is probable in designing the roof, the architect took

Fig. 78.

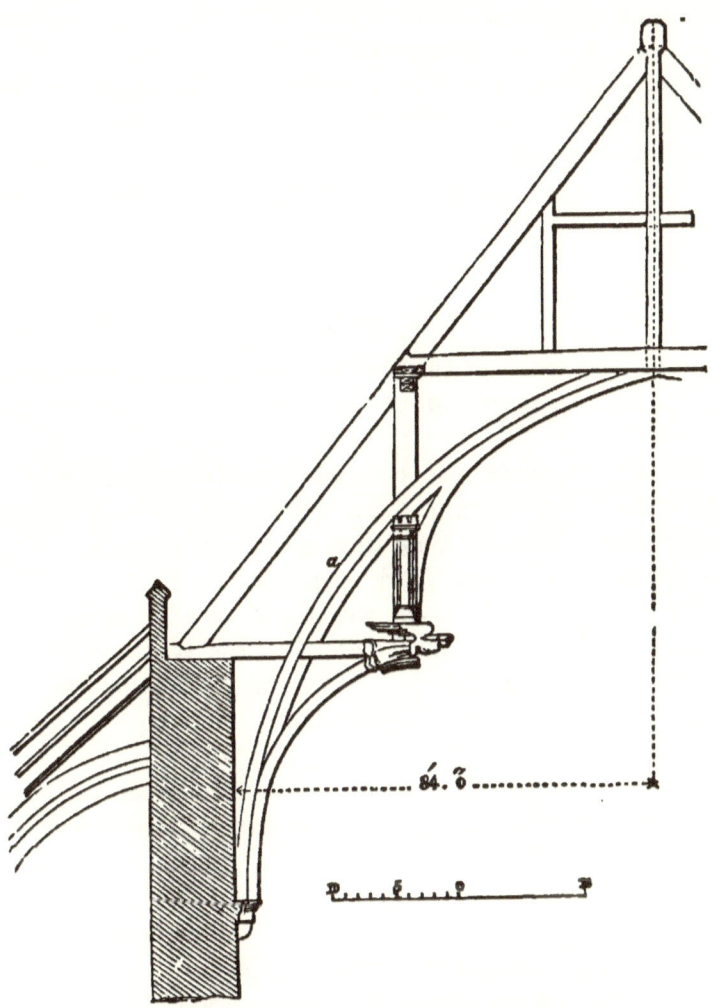

full advantage of the support afforded by the thickness of the walls and the buttresses; if, indeed, the latter were not added at the time the present roof was erected, in 1395. It has been ascertained that the weight of the roof rests on the top of the walls, the lower part of the arched rib only serving to distribute the thrust, and to assist in preventing the hammer beams from sliding on the walls.

D 3

113. The mediæval architects generally employed oak in the construction of their large roofs, the timbers being morticed and pinned together, as shown in fig. 65. This system of construction is impossible in fir and other soft woods, in which the fibres have little lateral cohesion, as the timber would split with the strain; and therefore, in modern practice, it is usual to secure the connections with iron straps or bolts passing round or through the whole thickness of the timbers.

ROOF COVERINGS.

114. The different varieties of roof coverings principally used may be classed under three heads: stone, wood, and metal.

Of the first class, the best kind is slate, which is used either sawn into slabs or split into thin laminæ. The different sizes of roofing slate in common use are given in the description of Slaters' Work, article 234.

In many parts of the country thin slabs of stone are used in the same way as roofing slate. In the Weald of Sussex the stone found in the locality is much used for this purpose, but it makes a heavy covering, and requires strong timbers to support it.

115. *Tiles* are of two kinds: *plain tiles*, which are quite flat; and *pantiles*, which are of a curved shape, and lap over each other at the sides. Each tile has a projecting ear on its upper edge, by which it is kept in its place. Sometimes plain tiles are pierced with two holes, through which oak pins are thrust for the same purpose.

116. Wooden coverings are little used at the present day, except for temporary purposes: *shingles* of split oak were formerly much used, and may still be seen on the roofs of some country churches.

117. *Metallic Coverings.*—The metals used for roof coverings are lead, zinc, copper, and iron.

118. Lead is one of the most valuable materials for this purpose on account of its malleability and durability, the

action of the atmosphere having no injurious effect upon
it. Lead is used for covering roofs in sheets weighing from
4 to 8 lbs. per sup. foot

119. Copper is used for covering roofs in thin sheets
weighing about 16 oz. per sup. foot, and from its lightness
and hardness has some advantages over lead; but the
expense of the metal effectually precludes its general
adoption.

120. Zinc has of late years superseded both lead and
copper to a considerable extent as roof coverings. It is used
in sheets weighing from 12 oz. to 20 oz. per sup. foot. It
is considered an inferior material to those just named; but
its lightness and cheapness are great recommendations,
and the manufacture has been much improved since its first
introduction.

121. Cast iron, coated with zinc to preserve it from rust-
ing, is now much used in a variety of forms. We have
already mentioned its adoption for covering the roofs of the
new Houses of Parliament. (See Note E, p. 157.)

122. All metallic coverings are subject to contraction
and expansion with the changes of the temperature, and
great care is requisite in joining the sheets to make them
lap over each other, so as to make the joints water-tight,
without preventing the play of the metal.

The following table of the comparative weights of different
roof coverings may be useful :—

	Cwt.	qrs.	lbs.
Plain tiles, per square of 100 ft. sup. . .	18	0	0
Pantiles	9	2	0
Slating, an average	7	0	0
Lead, 7 lb. to the sup. foot	6	2	0
Copper or zinc, 16 oz. do.	1	0	0

SUPPLY OF WATER.

123. The arrangements for distributing a supply of water
over the different parts of a building will depend very
materially on the nature of the supply, whether constant or
intermittent.

The most common method of supply from water-works is by pipes which communicate with private cisterns, into which the water is turned at stated intervals.

A cistern, in a dwelling-house, is always more or less an evil; it takes up a great deal of space, costs a great deal of money in the first instance, and often causes inconvenience, from leakage, from the bursting of the service pipes in frosty weather, and from the liability of the self-acting cock to get out of order

Fig. 74 shows the ordinary arrangements of a cistern for a dwelling-house. The common material for the cistern itself is wood lined with sheet lead; but slate cisterns have been much used of late. Large cisterns or tanks for the supply of breweries, manufactories, &c., are usually made of cast-iron plates, screwed together by means of flanges all round their edges.

Fig. 74.

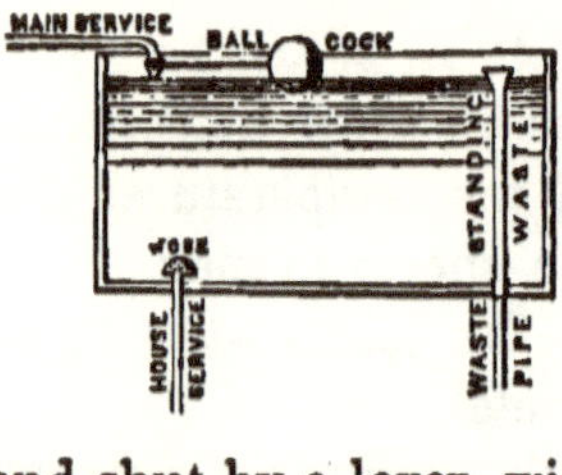

The service or feed pipe for a cistern, in the case of an intermittent supply, must be sufficiently large to allow of its filling during the time the water is turned on from the mains. The flow of water into the cistern is regulated by a *ball cock*, so called from its being opened and shut by a lever, with a copper ball, which floats on the surface of the water.

The service pipes to the different parts of the building are laid into the bottom of the cistern, but should not come within an inch of the actual bottom, in order that the sediment, which is always deposited in a greater or less degree, may not be disturbed: the mouth of each pipe should be covered by a *rose*, to prevent any foreign substances being washed into the pipes and choking the taps.

To afford a ready means of cleaning out the cistern, a waste pipe is inserted quite at the bottom, sufficiently large to draw off the whole contents in a short time when required; into this waste pipe is fitted a *standing waste*, which reaches nearly to the top of the cistern, and carries off the waste water, when, from any derangement in the working of the ball cock, the water continues running after the cistern is full   To prevent any leakage at the bottom of the standing waste, the latter terminates in a brass plug, which is ground to fit a washer inserted at the top of the waste pipe.

Where the supply of water is *constant*, instead of being intermittent, private cisterns may be altogether dispensed with; the main service pipes, not being required to discharge a large quantity of water in a short time, may be of smaller bore, and, consequently, cheaper, and a considerable length of pipe is saved, as the water can be laid on directly to the several taps, instead of having to be taken up to the cistern and then brought back again. The constant flow of water through the pipes also much diminishes the risk of their bursting in frosty weather from freezing of their contents.

WARMING AND VENTILATION.

124. The various contrivances employed for warming buildings may be classed as under :—

Methods of Warming independently of Ventilation.

1st. By close stoves, the heating surface being either of iron or of earthenware

2nd. By hot-air flues, passing under the floors.

3rd. By a system of endless piping heated by a current of hot water from a boiler, the circulation being caused by the cooling, and consequently greater weight, of the water in the lower or returning pipe.

Methods of Warming combined with Ventilation.

4th. By open fires placed in the several apartments.

5th. By causing air which has been previously heated to pass through the several rooms. This last system is more perfect than any of the others above described, both as regards economy of fuel and regulation of the temperature.

A great though common defect in the construction of fireplaces is their being placed too high; whence it is not unusual for the upper part of a room to be quite warm whilst there is a stratum of cold air next the floor, the effect of which is very injurious to health.

In all methods of warming, in which the air is heated by coming in contact with metallic heating surfaces, care should be taken that their temperature should not exceed 212°; as, when this limit is exceeded, the air becomes unfit for use, and offensive from the scorching of the particles of dust or other matters that are always floating in it.

125. There are two modes in which artificial ventilation is effected, each of which is very efficient.

The one most in use is to establish a draught in an air shaft or chimney communicating by flues with the apartments to be ventilated, the effect of which is to cause a constant current in the direction of the shaft, the air being admitted at the bottom of the building, and warmed or cooled as may be required, according to the season of the year.

The new House of Lords is ventilated in this manner The air is admitted at the bottom of the buildings, filtered by being passed through fine sieves, over which a stream of water is constantly flowing; warmed in cold weather by passing through steam cockles, and then, rising through the building, goes out through the roof into the furnace chimney, the draught being assisted by a steam jet from a boiler.

126. The other mode of ventilation to which we have alluded is on a completely opposite principle to that just described, the air being *forced into* the apartments by mechanical means, instead of being *drawn from* them by the draught in the chimney.

This latter plan is used with great success at the Reform Club House, the General Post-Office, and many other buildings, the air being thrown in by the action of large fans driven by steam power. (See Note F, p. 157.)

DRAINAGE.

127 This is a subject of equal importance with any of those previously noticed; but as two volumes of this series are devoted to its consideration, it is unnecessary to enter upon it in these pages.

SECTION II.

MATERIALS USED IN BUILDING.

128. The materials used in building may be classed under the following heads, viz.—

TIMBER.

STONE.—*See* volume on " Blasting Rocks and on Stone."

SLATE.—*See* Section IV. Art. 234.

BRICKS AND TILES.—*See* volume on " Brick-making and Tile-making."

LIMES AND CEMENTS.—*See* also Mr. Burnell's volume, in this series.

METALS.

GLASS.—*See* Section IV. Art. 258.

COLOURS AND VARNISHES.—*See* volume on " House-paint ing and Mixing Colours."

Some of these form the subjects of separate volumes of this series, to which the reader is therefore referred as above; and others are noticed in Section IV. of this volume Our remarks in this section will, therefore, be exclusively confined to the consideration of Timber, Limes and Cements, and Metals.

TIMBER.

129. If we examine a transverse section of the stem of a tree, we perceive it to consist of three distinct parts : the *bark*, the *wood*, and the *pith*. The wood appears disposed in rings round the pith, the outer rings being softer and containing more sap than those immediately round the pith which form what is called the *heart wood*.

These rings are also traversed by rays extending from the centre of the stem to the bark, called *medullary rays*.

The whole structure of a tree consists of minute vessels and cells, the former conveying the sap through the wood in its ascent, and through the bark to the leaves in its descent; and the latter performing the functions of secretion and nutrition during the life of the tree. The solid parts of a tree consist almost entirely of the fibrous parts composing the sides of the vessels and cells.

By numerous experiments it has been ascertained that the sap begins to ascend the spring of the year, through the minute vessels in the wood, and descends through the bark to the leaves, and, after passing through them, is deposited in an altered state between the bark and the last year's wood, forming a new layer of bark and sap wood, the old bark being pushed forward.

As the annual layers increase in number, the sapwood ceases to perform its original functions; the fluid parts are evaporated or absorbed by the new wood, and, the sides of the vessels being pressed together by the growth of the latter, the sapwood becomes heartwood or perfect wood, and until this change takes place it is unfit for the purposes of the builder.

The vessels in each layer of wood are largest on the side nearest the centre of the stem, and smallest at the outside. This arises from the first being formed in the spring, when vegetation is most active. The oblong cells which surround the vessels are filled with fluids in the early growth; but, as the tree increases in size, these become evaporated and absorbed, and the cells become partly filled with depositions of woody matter and indurated secretions, depending on the nature of the soil, and affecting the quality of the timber. Thus Honduras mahogany is full of black specks, while the Spanish is full of minute white particles, giving the wood the appearance of having been rubbed over with chalk. At a meeting of the Institution of Civil Engineers, March, 1842, it was stated by Professor Brande, that "a beech tree in Sir John Sebright's park in Hertfordshire, on being cut down, was found perfectly black all up the heart. On examination it was discovered that the tree had grown upon a mass of iron scoriæ from an ancient furnace, and that the wood had absorbed the salt of iron." This anecdote well explains the differences that exist between different specimens of the same kind of timber under different circumstances of growth; and it is probably the nature of the soil that causes the difference of character we have just named between Honduras and Spanish mahogany.

There is a great difference in the character of the annual rings in different kinds of trees. In some they are very distinct, the side next the heart being porous, and the other compact and hard, as in the oak, the ash, and the elm. In others the distinction between the rings is so small as scarcely to be distinguished, and the texture of the wood is nearly uniform, as in the beech and mahogany. A third class of trees have the annual rings very distinct and their pores filled with resinous matter, one part being hard and heavy, the other soft and light-coloured   All the resinous woods have this character, as larch, fir, pine, and cedar.

The medullary rings are scarcely perceptible to the naked eye in the majority of trees; but in some, as the oak and the beech, there are both large and small rings, which, when cut through obliquely, produce the beautiful flowered appearance called the silver grain.

130. In preparing timber for the uses of the builder there are three principal things to be attended to, viz. the age of the tree, the time of felling, and the seasoning for use.

131. If a tree be felled before it is of full age, whilst the heartwood is scarcely perfected, the timber will be of inferior quality, and, from the quantity of sap contained in it, will be very liable to decay. On the other hand, if the tree be allowed to stand until the heartwood begins to decay, the timber will be weak and brittle: the best timber

comes from trees that have nearly done growing, as there is then but little sapwood, and the heartwood is in the best condition.

132. The best time for felling trees is either in midwinter, when the sap has ceased to flow, or in midsummer, when the sap is temporarily expended in the production of leaves. An excellent plan is to bark the timber in the spring and fell it in winter, by which means the sapwood is dried up and hardened; but as the bark of most trees is valueless, the oak tree (whose bark is used in tanning) is almost the only one that will pay for being thus treated.

133. The seasoning of timber consists in the extraction or evaporation of the fluid parts, which are liable to decomposition on the cessation of the growth of the tree. This is usually effected by steeping the green timber in water, to dilute and wash out the sap as much as possible, and then drying it thoroughly by exposure to the air in an airy situation. The time required to season timber thoroughly in this manner will of course much depend on the sizes of the pieces to be seasoned; but for general purposes of carpentry, two years is the least that can be allowed, and, in seasoning timber for the use of the joiner, a much longer time is usually required.

134. *Decay of Timber.*—Properly seasoned timber, placed in a dry situation with a free circulation of air round it, is very durable, and has been known to last for several hundred years without apparent deterioration. This is not, however, the case when exposed to moisture, which is always more or less prejudicial to its durability.

When timber is constantly under water, the action of the water dissolves a portion of its substance, which is made apparent by its becoming covered with a coat of slime. If it be exposed to alternations of dryness and moisture, as in the case of piles in tidal waters, the dissolved parts being continually removed by evaporation and the action of the water, new surfaces are exposed, and the wood rapidly decays.

Where timber is exposed to heat and moisture, the albumen or gelatinous matter in the sapwood speedily putrefies and decomposes, causing what is called rot. The rot in timber is commonly divided into two kinds, the *wet* and the *dry*, but the chief difference between them is, that where the timber is exposed to the air, the gaseous products are freely evaporated; whilst, in a confined situation, they combine in a new form, viz. the dry-rot fungus, which, deriving its nourishment from the decaying timber, often grows to a length of many feet, spreading in every direction, and insinuating its delicate fibres even through the joints of brick walls.

In addition to the sources of decay above mentioned, timber placed in sea water is very liable to be completely destroyed by the perforations of the worm, unless protected by copper sheathing, the expense of which causes it to be seldom used for this purpose.

135. *Prevention of Decay.*—The best method of protecting woodwork from decay when exposed to the weather is to paint it thoroughly, so as to prevent its being affected by moisture. It is, however, most important not to apply paint to any woodwork which has not been thoroughly seasoned; for in this case the evaporation of the sap being prevented, it decomposes, and the wood rapidly decays.

Many plans have been proposed for the prevention of the rot. Kyan's process * consists in impregnating the timber with corrosive sublimate, thus converting the albumen into an indecomposable substance. This method, although not always successful, is undoubtedly of great use, particularly where inferior or imperfectly seasoned timber has to be used. It is, however, said to render the wood brittle.

Payne's process† consists in impregnating the wood with metallic oxides, alkalies, or earths, as may be required, and decomposing them in the wood, forming new and insoluble compounds. Timber thus prepared will not burn, but only smoulder

* Patented A.D. 1832.   † Patented A.D. 1841.

A process invented by Mr. Bethell,* and much used in railway works, is to impregnate the timber with oil of tar: this appears to be very successful in preventing decay, but the danger of accidents from fire is much increased. †

136. The variety of timber trees suitable to the purposes of the builder is very great; but fir and oak are the kinds chiefly used, although larch, beech, poplar and other woods, are employed to a limited extent in localities where they can be obtained more cheaply than foreign timber. Very little home-grown fir is used in England, as foreign timber, either in balks, or cut up into planks, deals, or battens, can be obtained at a moderate price in all the large towns in the kingdom, and is very superior to any grown in this country. Baltic timber is more esteemed than American, but a very great deal of the latter is used.

137. Fir is one of the most useful of the woods used by the builder. It is light, soft, easily worked, and very durable; but the lateral cohesion of the annual rings being very slight, it will not bear much strain, except in the direction of the length of the fibres. Red pine is also much used for carpenters' work, and is very durable. Yellow pine is sometimes used for joiners' work, but it is an inferior material, and liable to rot.

138. Oak, for purposes requiring strength, is preferred of English growth; that from Sussex is considered the best, being hard and fine-grained. The Dutch wainscot, which is grown in Germany, is a softer kind, and on that account not so apt to warp and twist, for which reason it is preferred to English oak for the purposes of the joiner: the texture of oak is very uniform, hard, and compact, which renders it superior to all other woods, as it will bear to be strained in any direction without fear of the rings separating, as in the resinous woods.

139. For internal finishings, mahogany is much used; that called Spanish, which comes from the West India Islands, is considered the best.

* Patented in 1838.   † See Note G, p. 158.

For joiners' and cabinet-makers' work, a great many kinds of fancy wood are imported, which are cut by machinery into thin slices, called *veneers*, and used as an ornamental covering to inferior work. In veneering, care should be taken that the body of the work be thoroughly seasoned, or it will shrink, and the veneer will fly off.

LIMES AND CEMENTS, MORTAR, ETC.

140. So much of the stability of brickwork and masonry depends upon the binding properties of the mortar or cement with which the materials are united, especially when exposed to a side pressure, as in the case of retaining walls, arches, and piers, that it is of no small importance to ascertain on what the strength of mortar really depends, and how far the proportions of the ingredients require modification, according to the quality of the lime that may have to be used.

It was long supposed that the hardness of any mortar depended upon the hardness of the limestone, from which the lime used in its composition was derived; but it was ascertained by the celebrated Smeaton, and since his time clearly shown by the researches of others, amongst whom may be named, Vicat in France, and Lieutenant-General Sir Charles Pasley in this country, that the hardness of the limestone has nothing to do with the matter, and that it is its chemical composition which regulates the quality of the mortar.

141. Limestone may be divided into three classes.

1st. Pure limes—as chalk.

2nd. Water limes—some of which are only slightly hydraulic, as the stone limes of the lower chalk, whilst others are eminently so, as the lias limes.

3rd. Water cements—as those of Sheppy and Harwich.

142. In making mortar the following processes are gone through.

1st. The limestone is calcined by exposure to strong

heat in a kiln, which drives off the carbonic acid gas con‑
tained in it, and reduces it to the state of *quick-lime.*

2nd. The quick-lime is *slaked* by pouring water upon
it, when it swells, more or less, with considerable heat,
and falls into a fine powder, forming a *hydrate* of lime.

3rd. The hydrate thus formed is mixed up into a
stiffish paste, with the addition of more water, and a proper proportion of sand, and is then ready for use.

143. *Pure Limes.*—*Chalk* is a pure carbonate of lime,
consisting of about 5 parts of lime combined with 4 of carbonic acid gas. It expands greatly in slaking, and will bear
from 3 to $3\frac{1}{2}$ parts of sand to one of lime, when made up
into mortar. Chalk lime mortar is, however, of little value,
as it *sets* or hardens very slowly, and in moist situations
never sets at all, but remains in a pulpy state, which renders it quite unfit for any work subjected to the action of
water, or even for the external walls of a building.

144. Gypsum, from which is made *plaster of Paris* for
cornices and internal decorations, is granular sulphate of
lime, and contains 26·5 of lime, 37·5 of sulphuric acid, and
17 of water. It slakes without swelling, with a moderate
heat, setting hard in a very short time, and will even set
under water; but as it is, like other pure limes, partly solu-
ble in water, it is not suitable for anything but internal
work

145. *Water limes* have obtained their name from the property they possess in a greater or less degree of setting
under water. They are composed of carbonate of lime,
mixed with silica, alumina, oxide of iron, and sometimes
other substances.

146. *Dorking lime*, obtained from the beds of the lower
chalk, at Dorking, in Surrey; and *Halling lime*, from a similar situation near Rochester, in Kent, are the principal limes
used in London for making mortar, and are slightly hydraulic; they expand considerably in slaking, but not so
much as the pure limes, and will make excellent mortar
when mixed with 3 parts of sand to 1 of lime. Mortar

made with these limes sets hard and moderately quick, and *when set*, may be exposed to considerable moisture without injury; but they will not set under water, and are therefore unfit for hydraulic works, unless combined with some other substance, as *puzzolana*, to give them water-setting properties.

147. The *blue lias limes* are the strongest water limes in this country. They slake very slowly, swelling but little in the process, and set very rapidly even under water; a few days only sufficing to make the mortar extremely hard. The lias limes will take a much smaller proportion of sand than the pure limes, the reason of which will be understood when it is remembered that they contain a considerable proportion of silica and alumina, combined with the lime in their natural state, and consequently the proportion of sand which makes good mortar with chalk lime, would ruin mortar made with Aberthaw, Watchet, Barrow, and other lias limes.

In the Vale of Belvoir, where the lias lime is extensively used, the common practice is to use equal parts of lime and sand for inside, and half sand to one of lime for face, work.

148. *Water Cements.*—These differ from the water-limes, as regards their chemical composition, only in containing less of carbonate of lime and more of silica and alumina. They require to be reduced to a fine powder after calcination, without which preparation they cannot be made to slake. The process of slaking is not accompanied by any increase of bulk, and they set under water in a short time, a few hours sufficing for a cement joint to become perfectly hard.

The principal supplies of cement-stone for the London market are derived from Harwich in Essex, and the Isle of Sheppy in Kent; where they are found in the London clay in the form of calcareous nodules.

Cement will not bear much sand without its cementitious properties being greatly weakened, the usual proportion being equal parts of sand and cement

149. The use of natural cement was introduced by Mr Parker, who first discovered the properties of the cement-stone in the Isle of Sheppy, and took out a patent for the sale of it in 1796, under the name of Roman cement.

Before that time, hydraulic mortar, for dock walls, harbour work, &c., was usually made, by mixing common lime with trass, from Andernach in Germany, or with puzzolana from Italy; both are considered to be volcanic products, the latter containing silica and alumina, with a small quantity of lime, potash, and magnesia. Iron is also associated with it in a magnetic state

150. The expense of natural puzzolana led to the manufacture of artificial puzzolana, which appears to have been used at an early date by the Romans, and has continued in use in the South of Europe to the present day; artificial puzzolana is made of pounded bricks or tile dust. The Dutch manufacture an artificial puzzolana from burnt clay, in imitation of the trass of Andernach, which is said to be a close imitation of the natural product.

151. The great and increasing demand for cement, and its great superiority for most purposes over lime mortar, have induced manufacturers to turn their attention to the manufacture of artificial cement, and this has been attended in many instances with perfect success; the artificial cements now offered for sale, formed by imitating the composition of the natural cement-stones, being mostly equal in quality, if not superior, to the Roman cement, the use of which has been partly superseded by them.

152. The quality of the *sand* used in making mortar is by no means unimportant. It should be clean and sharp; *i. e.* angular, and perfectly free from all impurities. The purer the lime the finer should be the quality of the sand, the pure limes requiring finer, and the cements a coarser sand, than the hydraulic limes.

CONCRETE AND BETON.

153. Rubble masonry, formed of small stones bedded in

mortar, appears to have been commonly used in England from an early period; and similar work, cemented witu hydraulic mortar, was constantly made use of by the Romans in their sea-works, of which many remains exist at tho present day in a perfectly sound state.

154. This mode of forming foundations, in situations where solid masonry would be inapplicable, has been re vived in modern times; in England under the name of concrete, and on the Continent under the name of béton. Although very similar in their nature and use, there are yet great differences between béton and concrete, which depend on the nature of the lime used, concrete being made with the weak water limes which will not set under water, whilst béton is invariably made with water-setting limes, or with limes rendered hydraulic by the addition of puzzolana. Describing the two by their differences, it may be observed that concrete is made with unslaked lime, and immediately thrown into the foundation pit; béton is allowed to stand before use, until the lime is thoroughly slaked : concrete is thrown into its place and rammed to consolidate it; béton is gently lowered and not afterwards disturbed: concrete must be thrown into a dry place, and not exposed to the action of water until thoroughly set; béton, on the contrary, is made use of principally *under water*, to save the trouble and expense of laying dry the bottom.

155. Concrete is usually made with gravel, sand, and ground unslaked lime, mixed together with water, the proportions of sand and lime being those which would make good mortar without the gravel, and, of course, varying according to the quality of the lime; with the common limes, slaking takes place at the time of mixing, and the quality of the concrete is all the better for the freshness of the lime. If lias lime be used, the concrete becomes béton, and must be treated accordingly

The lime in this case must be thoroughly slaked (which often takes many hours) before it can be considered fit for use; and, if this precaution be not attended to, the whole

E

of the work, after having set very hard on the surface, cracks and becomes a friable mass, from the slaking of the refractory particles after the body of the concrete has set.

The reader is referred, for further information on this subject, to the volume of this series on "Foundations and Concrete Works."

156. Asphalte, so much in use at the present day for foot-pavements, terrace-roofs, &c., is made by melting the asphalte rock, which is a carbonate of lime intimately combined with bitumen, and adding to it a small portion of mineral tar, which forms a compact semi-elastic solid, admirably adapted for resisting the effects of frost, heat, and wet.

Many artificial asphaltes have been brought under public notice from time to time, but they are all inferior to the natural asphalte, in the intimate combination of the lime and bitumen, which it appears impossible to effect thoroughly by artificial means.

METALS.

157. The metals used as building materials are iron, lead, copper, zinc, and tin.

158. *Iron.*—Iron is used by the builder in two different states, viz. cast iron and wrought iron, the differences between them depending on the proportion of carbon combined with the metal; cast iron containing the most, and wrought iron the least.

159. Previous to the middle of the last century, the smelting of iron was carried on with wood charcoal, and the ores used were chiefly from the secondary strata, although the clay ironstones of the coal measures were occasionally used.

The weald of Kent and Sussex* contained many ironworks during the seventeenth century. That at Lamberhurst, near Tunbridge Wells in Sussex, is noted as having

---

* The clay ironstones of Sussex are very rich, and are still raised in considerable quantities, and shipped for Wales and Newcastle.

furnished the cast-iron railing round St. Paul's Cathedral. The tilt hammers used in forging bar iron were chiefly worked by water power. A large pool in Beeding Forest, near Horsham in Sussex, still retains the name of the Hammer Pond, and the former sites of many old forges in the wealden district may still be traced by the heaps of cinders which yet remain here and there, and by the local names to which the works gave rise.

160. The introduction of smelting with pitcoal coke during the last century caused a complete revolution in the iron trade. The ores now chiefly used are the clay iron-stones of the coal measures, and the fuel, pitcoal or coke. Steam power is almost exclusively used for the production of the blast in the furnaces, and for working the forge hammers and rolling mills.

161. For the production of wrought iron in the ordinary manner, two distinct sets of processes are required. 1st. The extraction of the metal from the ore in the shape of cast iron. 2nd. The conversion of cast iron into malleable or bar-iron, by re-melting, puddling, and forging. The conversion of bar iron into steel is effected by placing it in contact with powdered charcoal in a furnace of cementation. (See Note H, p. 158.)

·162. *Cast iron* is produced by smelting the previously calcined ore in a blast furnace, with a portion of limestone as a flux, and pitcoal or coke as fuel. The melted metal sinks to the bottom of the furnace by its greater specific gravity. The limestone and other impurities float on the top of the melted mass, and are allowed to run off, forming *slag* or *cinder*. The melted metal is run off from the bottom of the furnace into moulds, where castings are required, and into furrows made in a level bed of sand, when the metal is required for conversion into malleable iron, the bars thus produced being called *pigs*.

163. In the year 1827, it was discovered that by the use of heated air for the blast, a great saving of fuel could be effected, as compared with the cold blast process.

E 2

The hot blast is now very extensively in use, and has the double advantage of requiring less fuel to bring down an equal quantity of metal, and of enabling the manufacturer to use raw pitcoal instead of coke, so that a saving is effected both in the quantity and cost of the fuel.

For a considerable time after its introduction it was held in great disrepute, which, however, may be chiefly attributed to the inferior quality of materials used, the power of the hot blast in reducing the most refractory ores offering a great temptation to obtain a much larger product from the furnace than was compatible with the good quality of the metal. The use of the hot blast by firms of acknowledged character has greatly tended to remove the prejudice against it; and in many iron works of high character, nothing but the hot blast with pitcoal is used in the smelting furnaces, the use of coke being confined to the subsequent processes.

Perhaps it may be laid down as a general principle, that where the pig iron is re-melted with coke in the cupola furnace, for the purposes of the ironfounder; or refined with coke in the conversion of forge pig into bar iron, it is of little consequence whether the reduction of the ore has been effected with the hot or the cold blast; but where castings have to be run directly from the smelting furnace, the quality of the metal will, no doubt, suffer from the use of the former.

164. Cast iron is divided by ironfounders into three qualities. No. 1, or *black cast iron*, is coarse-grained, soft, and not very tenacious. When re-melted it passes into No. 2, or *grey cast iron*. This is the best quality for castings requiring strength : it is more finely grained than No. 1, and is harder and more tenacious. When repeatedly re-melted it becomes excessively hard and brittle, and passes into No. 3, or *white cast iron*, which is only used for the commonest castings, as sash-weights, cannon-balls, and similar articles. White cast iron, if produced direct from the ore, is an indication of derangement in the working of the furnace, and

is unfit for the ordinary purposes of the founder, except to mix with other qualities. (See Note I, p. 159.)

165. Girders and similar solid articles are cast in sand moulds, enclosed in iron frames or *boxes*, each mould requiring an upper and lower box. A mould is formed by pressing sand firmly round a wooden *pattern*, which is afterwards removed, and the melted metal poured into the space thus left through apertures made for the purpose.

The moulds for ornamental work and for hollow castings are of a more complicated construction, which will be better understood from actual inspection at a foundry than from any written description.

Almost all irons are improved by admixture with others, and, therefore, where superior castings are required they should not be run direct from the smelting furnace, but the metal should be re-melted in a cupola furnace, which gives the opportunity of suiting the quality of the iron to its intended use. Thus, for delicate ornamental work, a soft and very fluid iron will be required, whilst, for girders and castings exposed to cross strain, the metal will require to be harder and more tenacious. For bed-plates and castings which have merely to sustain a compressing force, the chief point to be attended to is the hardness of the metal.

Castings should be allowed to remain in the sand until cool, as the quality of the metal is greatly injured by the rapid and irregular cooling which takes place from exposure to air if removed from the moulds in a red-hot state, which is sometimes done in small foundries to economise room

Staffordshire, Shropshire, and Derbyshire, afford the best irons for castings. The Scotch iron is much esteemed for hollow wares, and has a beautifully smooth surface, which may be noticed in the stoves and other articles cast by the Carron Company.

The Welsh iron is principally used for conversion into bar iron.

166. The conversion of forge pig into bar iron is effected by a variety of processes, which have for their object the

freeing the metal from the carbon and other impurities combined with it, so as to produce as nearly as possible the pure metal. We do not purpose to enter in these pages into any of the details of the manufacture of bar iron, or of its conversion into steel, as our business is rather with the ironfounder than the manufacturer; it may, however, be proper to state, that new processes have lately been patented, by which both malleable iron and steel may be produced directly from the ore, without the use of the smelting furnace, a plan which is likely to be attended with beneficial results, both as regards economy and quality of metal.

167. *Lead.*— Lead is used by the mason for securing dowels, coating iron cramps, and similar purposes, *see* Sec-tion IV., Plumber.

Lead is also used by the smith in fixing iron railings, and other work where iron is let into stone; but the use of lead in contact with iron is always to be avoided, if pos-sible, as it has an injurious effect upon the latter metal, the part in contact with the lead becoming gradually softened.

The chief value of lead, however, to the builder, is as a covering for roofs, and for lining gutters, cisterns, &c., for which uses it is superior to any other metal. For these purposes the lead is cast into sheets, and then passed be-tween rollers in a *flatting-mill*, until it has been reduced to the required thickness.

Cast-lead is often made by plumbers themselves from old lead taken in exchange; but it is very inferior to the *milled lead* of the manufacturer, being not so compact, and often containing small air-holes, which render it unfit for any but inferior purposes

168. *Copper.*—*See* Section IV., Coppersmith.

169. *Zinc.*—*See* Section IV., Zincworker.

170. *Brass* is an alloy of copper and zinc, the best proportions being nearly two parts of copper to one of zinc.

171. *Bronze* is a compound metal, composed of copper and tin, to which are sometimes added a little zinc and lead.

The best proportions for casting statues and bas-reliefs appear to be attained when the tin forms about 10 per cent. of the alloy.

By alloying copper with tin, a more fusible metal is obtained, and the alloy is much harder than pure copper; but considerable management is required to prevent the copper from becoming refined in the process of melting, a result which has frequently happened to inexperienced founders.

172. *Bell-metal* is composed of copper and tin, in the proportion of 78 per cent. of the former to 22 per cent. of the latter

---

SECTION III.

STRENGTH OF MATERIALS.

173. There are three principal actions to which the materials of a building are exposed.

1st. *Compression*—as in the case of the stones in a wall.

2nd *Tension*—as in the case of a king-post or tie-beam.

3rd. *Cross strain*—as in the case of a bressummer, floor-joists, &c.

The last of the three is the only one against which precautions are especially necessary, as in all ordinary cases the resistance of the materials used for building is far beyond any direct crushing or pulling force that is likely to be brought upon them.

174. 1st. *Resistance to Compression.*—The following table shows the force required to crush $1\frac{1}{2}$ in. cubes of several kinds of building material.—

	lbs.		lbs.
Good brick . . .	1817	Portland stone	10,284
Derbyshire grit .	7070	Granite ,,	14,300

These amounts so far exceed any weight that could have to be borne on an equal area, under ordinary circumstances, that it is quite unnecessary in the erection of a building to make any calculations on this head when using these or similar materials.

Cast iron may be considered as practicably incompressible ; *wrought iron* may be flattened under great pressure, but cannot be crushed. *Timber* may be considered, for practical purposes, as nearly incompressible, when the weight is applied in the direction of the fibres, as in the case of a wooden story-post ; but the softer kinds, as fir, offer little resistance, when the weight is applied at right angles to the fibres, as in the case of the sill of a partition ; and, besides this, timber, however well-seasoned, will al ways shrink, more or less, in the direction of its thickness, so that no important bearings should be trusted to it.

175. *2nd. Resistance to Tension.*—The principal building materials that are required to resist direct tension are *timber* and *wrought iron.* (See Note K, p. 159.)

The following table shows the weight in tons required to tear asunder bars 1-inch square of the following materials :—

	Tons.
Oak	$5\frac{1}{6}$
Fir	$5\frac{1}{4}$
Cast iron	$7\frac{3}{4}$
Wrought iron . . .	10
Wrought copper	15
English bar iron . . .	25
Blistered steel	$59\frac{1}{2}$

Cast iron, however, although included in the above table, is an unsuitable material for the purpose of resisting tension, being comparatively brittle. With regard to *timber*, it is practically impossible to tear asunder a piece of even moderate size, by a force applied in the direction of the fibres, and therefore the dimensions of king-posts, tie-beams, and other timbers which have to resist a pulling force, are regulated by the necessity of forming proper joints and connections with the other parts of the framing to which they belong, rather than by their cohesive strength But it must be borne in mind, that although the strength of all kinds of timber is very great in the direction of the fibres, the lateral cohesion of the annual rings is in many kinds of wood very slight, and must be assisted by iron straps in all doubtful cases. The architects of the middle ages executed their magnificent wooden roofs without these aids, but they worked in oak, and not in soft fir, which would split and rend if treated in the same way.

Wrought iron is extensively used for bolts, straps, tie-rods, and all purposes which require great strength, with small sectional area; one-fourth of the breaking weight is usually said to be the limit to which it should be strained; but, in all probability, this amount might be doubled without any injurious effects.

STRENGTH OF BEAMS.

176. 3rd. *Cross Strain.*—In calculating the strength of beams when exposed to cross or transverse strain, two principal considerations present themselves: 1st, The mechanical effect which any given load will produce under varying conditions of support; and 2ndly, The resistance of the beam, and the manner in which this is affected by the form of its section.

177. 1st. *Mechanical Effect of a given Load under varying Circumstances.*—If a rectangular beam be supported at each end and loaded in the middle, the strength of the beam, its section remaining the same, will be inversely as the distance between the supports, the weight acting with a leverage which increases at this distance.* If a beam be

* It may be as well to observe that, although this is true as to the strength of beams under ordinary circumstances, it does not hold good when

E 3

fixed at one end and weighted at the other (fig. 75), its strength will be half that of a similar beam of double the

Fig. 75.

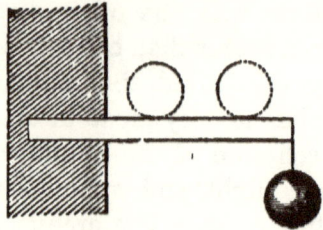

length supported as first described (fig 76). A parallel case to this is that of a beam supported in the middle and

Fig. 76.

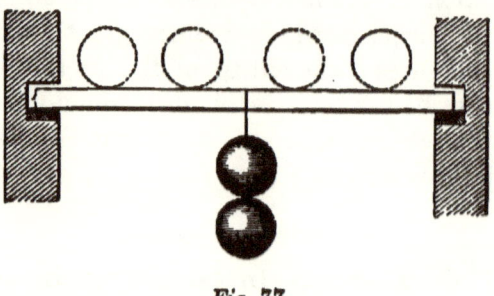

Fig. 77.

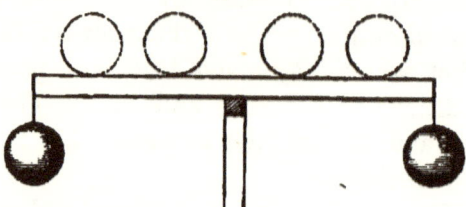

loaded at the ends (fig 77). In each of the above cases the beam will bear double the load if it be equally distributed over its whole length, as shown by the dotted lines;

the loading is carried to the breaking point, the deflection of the beam causing an increase or diminution of the leverage according to the mode of support. The difference of strength arising from this cause is, however, too trifling to be taken into consideration, except in delicate experiments on the ultimate strength of beams.

and lastly, the strength of a beam firmly fixed at the ends is to its strength when loosely laid on supports as 3 to 2 (*see* fig. 78).

Fig. 78.

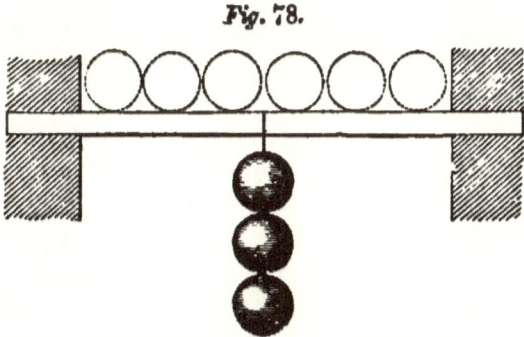

These results may be simply expressed thus :

Let *s* be the weight which would break a beam of given length and scantling fixed at one end and loaded at the other ;

then 2 *s* would break the same beam fixed at one end and uniformly loaded ;

    4 *s* would break the same beam supported at each end and loaded in the middle ;

    6 *s* would break the same beam fixed at each end and loaded in the middle ;

    8 *s* would break the same beam supported at each end and uniformly loaded ;

    12 *s* would break the same beam fixed at each end and uniformly loaded.

178. 2nd. *Resistance of the Beam.*—If a beam be loaded so as to produce fracture, this will take place about a centre or neutral axis, below which the fibres will be *torn* asunder, and above which they will be crushed. This may be very clearly illustrated by drawing a number of parallel lines with a soft pencil on the edge of a piece of India rubber, and bending it round, when it will be seen that the lines are brought closer together on the concave, and stretched further asunder on the convex side, whilst, between the two edges, a neutral line may be traced, on which the divisions

remain of the original size, which neutral line divides the fibres that are subjected to compression from those in a state of tension (*see* fig. 79).

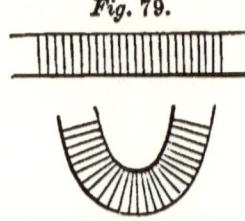

Fig. 79.   The resistance of a rectangular beam will, therefore, depend, 1st, on the number of fibres, which will be proportionate to its breadth and depth; 2nd, on the distance of those fibres from the neutral axis, and the consequent leverage with which they act, which will also be as the depth; and, lastly, on the actual strength of the fibres, which will vary with different materials, and can only be determined approximately from actual experiments on rectangular beams of the same material as those whose strength is required to be estimated.

The actual strength of any rectangular beam will, therefore, be directly as its breadth multiplied by the square of the depth, and inversely as its length; or, calling s the transverse strength of the material, as in art. 177, b the breadth, d the depth, l the length between the supports, and W the breaking weight,

$$W = \frac{s\,b\,d^2}{l}$$

The following may be taken as the value of s for iron and timber, the length being taken in feet, the breadth and depth in inches, and the breaking weight in pounds.

	Constant multiplier for rectangular beams fixed at one end and loaded at the other.		Constant multiplier for rectangular beams loosely supported at the ends and loaded in the middle.
Wrought iron	512		2048
Cast ditto	500	× 4	2000*
Fir and English oak	100		400

* The above is an average value calculated from a great number of published experiments on different irons. The best Derbyshire and Staffordshire irons might probably be taken as high as 2500, whilst the ordinary Scotch hot-blast irons could not be trusted to bear more than from 1700 to 1000.

It must be remembered that the numbers here given indicate the breaking weight, not more than one-third of which should ever be applied in practice. Timber is permanently injured if more than even one-fourth of the breaking weight is placed on it, and, therefore, this limit should never be passed.

A single example will suffice to show the importance of the principles just explained, and the lamentable results that may follow from ignorance of them. If we take a fir binding joist, say 9 in. × 4 in., which is to have a bearing of 12 ft. between its supports, and place it edgeways, it will require to break it a weight $= \dfrac{400 \times 4 \times 9^2}{12} = 10{,}800$ lbs.: but if, for the purpose of gaining height, we place it flatways, it will break with a weight $= \dfrac{400 \times 9 \times 4^2}{12} = 4800$ lbs., or less than one-half.

179. We may see from this example that the shape of any beam has a great influence on its strength ; and in making beams of iron, which can be cast with great facility in any required shape, it becomes an important question how to obtain the strongest form of section with the least expenditure of metal.

The usual section given to cast-iron girders is that of a thin and deep rectangular beam, with flanges or projections on each side at top and bottom ; where the strength of the metal will be most effective, as being at the greatest possible distance from the neutral axis (fig. 80).

Fig. 80.

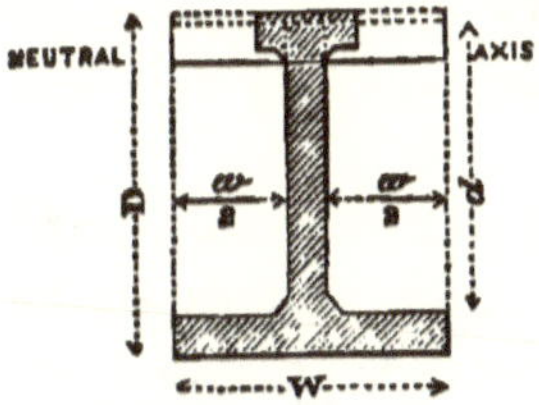

The great question now is, what should be the relative thickness of the top and bottom flanges, the centre part of the beam having been made as thin as is consistent with sound casting ?

If the metal were incompressible, the top flanges might be infinitely thin ; if incapable of extension, the bottom ones might be indefinitely reduced. If it offered equal resistance to tension and compression, the neutral axis would occupy the centre of the beam, and the top and bottom flanges would require to be of equal strength.

We are indebted to Mr. Eaton Hodgkinson for the publication * of a valuable set of experiments conducted by him, having for their object the determination of the position of the neutral axis in cast-iron beams. The result of his experiments is, that in cast-iron rectangular beams, the position of the neutral axis at the time of fracture is at about one-seventh of the whole depth of the beam below its upper surface Hence, in girders with flanges, the thickness of the bottom flanges should be six times that of the upper ones (supposing them to be of the same width), in order to obtain the greatest strength with the least metal. Practically it would be almost impossible to cast a beam thus proportioned, and, therefore, the top flanges are made of the same thickness, or nearly so, as the bottom ones, but of a less width, so as to contain the same relative quantity of metal, disposed in a more con venient form for casting (fig. 80).

The difficulty of making sound castings where the parts are of unequal thickness also renders it necessary to make the thickness of the middle rib nearly equal to that of the flanges.

180. To calculate the strength of a cast-iron beam, the sectional area of whose top flanges is $\frac{1}{6}$ of that of the bottom ones, we must find that of a rectangular beam of the same extreme depth and width, and deduct from it the resistance of the portions omitted between the top and bottom flanges (fig. 80).

If we call the whole width of the bottom of the beam, W, the sum of the widths of the two bottom flanges, w, the whole depth of the beam, D, and the vertical distance between the flanges d (on the supposition that the top flanges

* Experimental Researches on the Strength and other Properties of Cast Iron. 8vo, 1846. WEALE.

are of the same widths as the bottom ones, and $\frac{1}{6}$ of their thickness, as shown by the dotted lines in fig. 80), the distance between the supports, l, the strength of the material, s, as in art. 177, and if the weight required to break a beam when loosely supported at the ends and loaded in the middle be called x,

$$\text{Then } x = \frac{(W D^2 - w d^2) 4 s}{l},$$

and if we take the length in feet and the other dimensions in inches, and call $s = 560$ lbs., which is not too much for the best Staffordshire irons; then

$$4 s = 2240 \text{ lbs.} = 1 \text{ ton; and therefore } \frac{W D^2 - w d^2}{l} =$$

breaking weight in tons.

The value of d in this rule will be $D - \frac{1}{6}$ of the thickness of the bottom flanges, and so long as the sectional area of the top flanges is more than $\frac{1}{6}$ of that of the bottom ones,[*] the rule may be applied to girders of variously proportioned flanges, as the additional strength gained by increasing the size of the top flanges beyond the proportion here named is very small in proportion to the metal used, and, in neglecting to take it into account, we are sure to err on the safe side

181. It must not be supposed, that because increasing the thickness of the top flanges does not materially increase the resistance to vertical pressure, it is on that account useless: on the contrary, where a beam is of considerable depth in proportion to the widths of the bottom flanges, it will often be desirable to make the top flanges more than $\frac{1}{6}$ of the bottom ones, in order to prevent the girder from twisting laterally, and to increase the resistance to any side

---

[*] It must be remembered that in making the top flanges narrower than the bottom ones for convenience of casting, as the bulk of the metal is brought nearer to the neutral axis by so doing, the sectional area of the top flanges must be rather more than $\frac{1}{6}$ of that of the bottom ones, in order to keep the position of the neutral axis the same as in a rectangular beam.

thrust to which it may be exposed from brick arches or otherwise.

182. In practice, it is not desirable to load iron girders beyond ¼ of their ultimate strength, and they should be *proved* before use by loading them to this extent or a little more, but care should be taken never to let the proof exceed ½ the breaking weight, as a greater load than this strains and distresses the metal, making it permanently weaker. The ultimate strength of a girder of the usual proportions may be approximately ascertained from its deflexion under proof on the assumption that a load equal to half the breaking weight will cause a deflection of $\frac{1}{480}$ of its length.*

183. *Trussed Timber Beams.*—Timbers exposed to severe strain require to be *trussed* with iron, and this may be done in two ways: 1st, by inserting cast-iron struts, as in fig. 81,

Fig. 81.

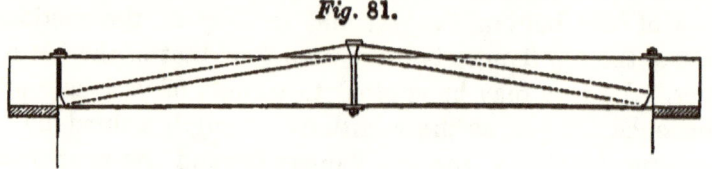

thus placing the whole, or nearly the whole, of the woodwork in a state of tension; 2nd, by wrought-iron tension rods, as in fig. 82, which take the whole of the tension,

Fig. 82.

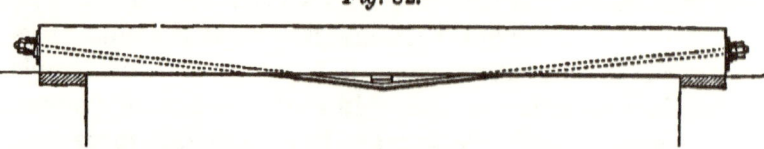

whilst the timber is thrown entirely into compression The latter mode of trussing is now very extensively used in strengthening the carriages of travelling cranes and for similar purposes; and, by its use, a balk of timber which will barely support its own weight safely without assistance,

* The author is indebted for this rule to the manager of the Phœnix Foundry, Derby. See also Note L. p. 160.

may be made to carry a load of many tons without sensible deflection.

STRENGTH OF STORY-POSTS AND CAST-IRON PILLARS.

184. When a piece of timber, whose length is not less than 8 or 10 times its diameter, is compressed in the direction of its length, as in the case of a wooden story-post supporting a bressummer, it will give way if loaded beyond a certain point, not by crushing, but by bending, and will ultimately be destroyed by the cross strain, just as a horizontal beam would be by vertical pressure applied at right angles to the fibres. The rules for determining the dimensions of a piece of timber to support a given weight without sensible flexure are very complicated, and are of little practical value, as they depend upon the condition that the pressure is exactly in the direction of the axis of the post—a condition rarely fulfilled in practice.

185. Wooden story-posts have been to a great extent superseded by the use of cast-iron pillars, which possess great strength with a small sectional area, and are on that account particularly well adapted to situations where it is of consequence to avoid obstructing light, as in shop-fronts.

In determining the design of a cast-iron pillar, whose length is 20 or 30 times its diameter, two points have to be considered: 1st, the liability to flexure; 2nd, the risk of the ends being crushed by the load not acting in the direction of the axis of the pillar.

Fig. 83.

The resistance to flexure is greatly increased by enlarging the bearing surface at the ends of the pillar, as in fig. 83, which, on the other hand, increases the liability of the ends to fracture, in the event of the load being thrown on the side instead of on the centre of the column, by any irregular settlement of the building. The judicious architect will, therefore, take a mean course, swelling out the capitals and bases of his cast-iron pillars enough to prevent their shafts from bend-

ing, but at the same time avoiding any thin flanges or projections, which might be liable to be broken. No theoretical rule for determining the proportions of a cast-iron
pillar depending on the weight to be supported can be depended on in practice. The real measure of the strength
of a cast-iron story-post must be the power of resisting
any lateral force which may be brought against it; and as
a slight side blow will suffice to fracture a pillar which is
capable of supporting a vertical pressure of very many tons,
we have only to make sure of the lateral strength, and we
are quite certain to be on the safe side as regards any
vertical pressure which it may have to sustain.

186. Besides the above cases of transverse strain, there
are others arising from irregular settlements, which are
amongst the greatest difficulties with which the builder has
to contend. Thus, to take a familiar instance, the window
sills of a dwelling-house are often broken by the settlement
of the brickwork being greater in the piers than under the
sills, from the greater pressure on the mortar joints; and
this will take place with a difference of settlement which
can scarcely be detected, even by careful measurement.*
We need not here enlarge on this subject, as we have several
times in the preceding pages had occasion to notice both
the causes of irregular settlement, and the precautions to be
taken for its prevention.

The strength of materials to resist *torsion* or twisting, as
in the case of a driving shaft, is an important consideration
in the construction of machinery, but is of little consequence in the erection of buildings, and therefore need not
be noticed in these pages.

* The reader need scarcely be told that a careful builder will always defer
pinning up his sills until some time has been allowed for the settlement of
the brickwork, but this will not always prevent ultimate fracture.

SECTION IV.
USE OF MATERIALS.

EXCAVATOR

187. The digging required for the foundations of com·
mon buildings usually forms part of the business of the
bricklayer, and is paid for at per cubic yard, according to
the depth of the excavation, and the distance to which the
earth has to be wheeled; this being estimated by the *run*
of 20 yards

In large works, which require coffer-dams and pumping
apparatus to be put down before the ground can be got out
for the foundations, the work assumes a different character,
and is paid for accordingly; the actual excavation being only
a small item of the total cost compared with those of dredg·
ing, piling, puddling, shoring, pumping, &c.

The workmen required for the construction of coffer-dams
and similar works are labourers of a superior class, accustomed to the management of pile-engines and tackle, and
competent to the execution of such rough carpenter's work
as is required in timbering large excavations.

The methods in use of constructing coffer-dams, driving
piles, and executing other work connected with foundations,
are described in the volume of this series on "Foundations
and Concrete Works;" to which the reader is referred for
further information on the subject

BRICKLAYER

188. The business of a bricklayer consists in the execution of all kinds of work in which brick is the principal
material; and in London it always includes tiling and pav·
ing with bricks or tiles. Where undressed stone is much
used as a building material, the bricklayer executes this
kind of work also, and in the country, the business of the
plasterer is often united with the above-named branches

189. The tools of the bricklayer are the *trowel* to take up

and spread the mortar, and to cut bricks to the requisite length : the *brick axe*, for shaping bricks to any required bevel ; the *tin saw*, for making incisions in bricks to be cut with the axe, and a *rubbing-stone*, on which to rub the bricks smooth after being roughly axed into shape. The *jointer* and the *jointing-rule* are used for *running* the centres of the mortar-joints The *raker*, for raking out the mortar from the joints of old brickwork previous to re-pointing. The *hammer*, for cutting chases and splays. The *banker* is a piece of timber about 6 feet long, raised on supports to a convenient height to form a table on which to cut the bricks to any required gauge, for which *moulds* and *bevels* are required. The *crowbar*, *pick-axe*, and *shovel* are used in digging out the foundations, and the *rammer* in punning the ground round the footings, and in rendering the foundation firm where it is soft by beating or ramming.

To set out the work and to keep it true, the bricklayer uses the *square*, the *level*, and the *plumb-rule;* for circular or battering work he uses *templets* and *battering-rules ; lines* and *pins* are used to lay the courses by; and *measuring-rods* to take dimensions. When brickwork has to be carried up in conjunction with stonework, the height of each course must be marked on a *gauge-rod*, that the joints of each may coincide.

190. The bricklayer is supplied with bricks and mortar by a labourer, who carries them in a *hod.* The labourer also makes the mortar, and builds and strikes the scaffolding.

191. The bricklayer's scaffold is constructed with *standards, ledgers,* and *putlogs.* The standards are fir poles, from 40 to 50 ft. long, and 6 or 7 in. diameter at the butt ends, which are firmly bedded in the ground. When one pole is not sufficiently long, two are lashed together, top and butt, the lashings being tightened with wedges. The ledgers are horizontal poles placed parallel to the walls, and lashed to the standards for the support of the putlogs. The putlogs are cross pieces usually made of birch, and about 6 ft. long, one end resting in the wall, the

other on a ledger. On the putlogs are placed the scaffold boards, which are stout boards hooped at the ends to prevent them from splitting.

192. A bricklayer and his labourer will lay in a single day about 1000 bricks, or about two cubic yards

193. The tools required for tiling are—the *lathing-hammer*, with two gauge marks on it, one at 7, and the other at 7½ inches; the *iron lathing staff*, to clinch the nails; the *trowel*, which is longer and narrower than that used for brickwork; the *bosse*, for holding mortar and tiles, with an iron hook to hang it to the laths or to a ladder; and the *striker*, a piece of lath about 10 in. long, for clearing off the superfluous mortar at the feet of the tiles.

194. Brickwork is measured and valued by the rod, or by the cubic yard, the price including the erection and use of scaffolding, but not centering to arches, which is an extra charge.

Bricknogging, pavings, and facings, by the superficial yard.

Digging and steining of wells and cesspools by the foot in depth, according to size, the price increasing with the depth.

Plain tiling and pantiling are valued per square of 100 feet superficial.

A journeyman bricklayer receives from 4*s.* to 5*s.* 6*d.*, and a labourer from 2*s.* 6*d.* to 3*s.* 6*d* a day.

The following memoranda may be useful:—

Weight of different kinds of Earth.

13 cubic feet of chalk weigh one ton.
17 ,, clay ,,
18 ,, nightsoil ,,
21¾ ,, gravel ,,
23½ ,, sand ,,

Nightsoil is removed in carts containing 45 cubic feet, or 2½ tons.

Twenty-seven cubic feet or 1 cubic yard is called a single load. and 2 cubic yards a double load.

A measure of lime is 27 cubic feet and contains 21 striked bushels.

A bricklayer's hod measures 1 ft. 4 in. × 9 in. × 9 in., and contains 20 bricks.

A rod of brickwork measures 16½ ft. square, 1½ brick thick (which is called the reduced or standard thickness), or 272 ft. 3 in. superficial, or 306 cubic feet, or 11⅓ cubic yards.

Table of the Sizes and Weights of various Articles.

DESCRIPTION.	Length.		Breadth.		Thickness.		Weight.	
	ft.	in.	ft.	in.	ft.	in.	lbs.	oz.
Stock bricks . . each	0	8¾	0	4¼	0	2½	5	0
Paving do. . . . . "	0	9	0	4½	0	1¾	4	0
Dutch clinkers . . "	0	6¼	0	3	0	1½	1	8
12-in. paving tiles . "	0	11¾	0	11¾	0	1½	13	0
10-in. do. . . . "	0	9¾	0	9¾	0	1	8	9
Pantiles . . . . "	1	1½	0	9½	0	0½	5	4
Plain tiles . . . . "	0	10½	0	6½	0	0⅝	2	5
Pantile laths per 10 ft. bundle	120	0	0	1½	0	1	4	6
Do. " 12 ft. do.	144	0	0	1½	0	1	5	0
N.B.—A bundle contains twelve laths.								
Plain tile lathes per bundle .	500	0	0	1	0	0¼	3	0
N.B.—Thirty bundles of laths make a load.								

A rod of brickwork, laid four courses to a foot in height, requires 4353 stock bricks.

Ditto, 11¼ in. to 4 courses, 4533 stock bricks.

These calculations are made without allowing for waste, which is unnecessary, because the space occupied by flues, bond timber, &c., and for which no deduction is made, more than compensates for any waste; and in building dwelling-houses, 4300 stocks to a rod is sufficient.

If laid dry, 5370 stocks to the rod.

4900 ditto, in wells and circular cesspools.

A rod of brickwork, laid 4 courses to gauge 12 in., contains 235 cubic feet of bricks and 71 cubic feet of mortar, and weighs about 15 tons.

A rod of brickwork requires 1¼ cubic yard of chalk lime and 3 single loads of sand, or 1 cubic yard of stone lime and 3¼ loads of sand, or 36 bushels of cement and an equal quantity of sharp sand.

A cubic yard of mortar requires 9 bushels of lime and 1 load of sand.

Lime and sand, and likewise cement and sand, lose ⅓ of their bulk when made into mortar.

The proportion of mortar or cement, when made up, to the lime or cement and sand before made up, is as 2 to 3.

Lime or cement and sand to make mortar require as much water as is equal to ⅓ of their bulk.

A cubic yard of concrete requires 34 cubic feet of material; or, if the gravel is to the lime as 6 to 1, a cubic yard of concrete will require 1·1 cubic yard of gravel and sand and 3 bushels of lime.

Facing requires 7 bricks per foot superficial

Gauged arches, 10 ditto ditto.

Bricknogging per yard superficial requires 30 bricks on edge, or 45 laid flat.

195. *Paving :—*

Stock bricks laid flat require 36 per yard superficial.

Ditto on edge	,,	52	,,
Paving bricks laid flat	,,	36	,,
Ditto on edge	,,	82	,,
Dutch clinkers ditto	,,	140	,,
12-inch paving tiles	,,	9	,,
10-inch ditto	,,	13	,,

196. *Tiling :—*

Description.	Gauge in Inches.	No. required per square.
With pantiles	12	150
Ditto	11	164
Ditto	10	180

N.B.—A square of pantiling requires 1 bundle of laths and 1¼ hundred of sixpenny nails.

Description.	Gauge in Inches.	No. required per square.
With plain tiles	4	600
Ditto	3½	700
Ditto	3	80

N.B A square of plain tiling requires 1 bundle of laths, 1 peck of tile pins, and 3 hods of mortar

| Plain tiles laid flat . . . . | | 210 |

MASON.

197 The business of the mason consists in *working* the stones to be used in a building to their required shape, and in *setting* them in their places in the work. Connected with the trade of the mason are those of the *Stonecutter*, who *hews* and cuts large stones roughly into shape preparatory to their being *worked* by the mason, and of the *Carver*, who executes the ornamental portions of the stone-work of a building, as enriched cornices, capitals, &c.

198. Where the value of stone is considerable, it is sent from the quarry to the building in large blocks, and cut into slabs and scantlings of the required size with a stone-mason's saw, which differs from that used in any other trade in having no teeth. It is a long thin plate of steel, slightly jagged on the bottom edge, and fixed in a frame; and, being drawn backwards and forwards in a horizontal position, cuts the stone by its own weight. To facilitate the operation, a heap of sharp sand is placed on an inclined plane over the stone, and water allowed to trickle through it, so as to wash the sand into the saw-cut. Of late years machinery worked by steam-power has been used for sawing marble into slabs to a very great extent, and has almost entirely superseded manual labour in this part of the manufacture of chimney-pieces.

Some freestones, as Bath-stone, are so soft as to be easily cut with a toothed saw worked backwards and forwards by two persons.

The harder kinds of stones, as granites and gritstones, are brought roughly into shape at the quarry, with an axe or a scappling hammer, and are then said to be *scappled.*

199. The tools used by the mason for cutting stone consists of the *mallet* and *chisels* of various sizes. The mason's mallet differs from that used by any other artisan, being similar to a dome in contour, excepting a portion of the broadest part, which is rather cylindrical; the handle is short, being only sufficiently long to enable it to be firmly grasped.

In London the tools used to work the faces of stone are the *point*, which is the smallest description of chisel, being never more than a quarter of an inch broad on the cutting edge; the *inch tool;* the *boaster*, which is 2 in. wide; and the *broad tool*, of which the cutting edge is $3\frac{1}{2}$ in. wide. The tools used in working mouldings and in carving are of various sizes, according to the nature of the work.

Besides the above cutting tools the mason uses the *banker* or bench, on which he places his stone for convenience of working, and *straight edges, squares, bevels,* and *templets* for marking the shapes of the blocks, and for trying the surfaces as the work proceeds. Any angle greater or less than a right angle is called a bevel angle, and a *bevel* is formed by nailing two straight edges together at the required angle; a *bevel square* is a square with a shifting stock which can be set to any required bevel. A templet is a pattern for cutting a block to any particular shape; when the work is moulded the templet is called a *mould.* Moulds are commonly made of sheet zinc, carefully cut to the profile of the mouldings with shears and files.

For setting his work in place the mason uses the *trowel, lines, and pins,* the *square* and *level,* and *plumb* and *battering rules,* for adjusting the faces of upright and battering walls.

200. The mason's scaffold is double, that is, formed with two rows of standards, so as to be totally independent of the walls for support, as putlog holes are inadmissible in masonry.

F

During the last ten years the construction of scaffolds, with round poles lashed with cords has been entirely super-seded in large works by a system of scaffolding of square timbers connected by bolts and dog irons.

The hoisting of the materials is performed from these scaffolds by means of a travelling crane, which consists of a double travelling carriage running on a tramway formed on stout sills laid on the top of two parallel rows of standards. The crab-winch is placed on the upper carriage, and, by means of the double motion of the two carriages, can be brought with great ease and precision over any part of the work lying between the two rows of standards

The facilities which are afforded by these scaffolds and travelling cranes for moving heavy weights over large areas, have led to their extensive adoption, not only in the erec-tion of buildings, but on landing wharfs, masons and iron-founders' yards, and similar situations, where a great saving of time and labour is effected by their use

Scaffolding of square timbers appears to have been little used in England before A.D. 1837, when Messrs. Cubitt, of Gray's Inn Road, applied it to the erection of the entrance gateway of the Euston station of the North-Western Rail-way. Since then it has been very generally used in large works, amongst which may be mentioned the Reform Club House, in A.D. 1838, and the Nelson Column, commenced A.D 1840, where it was carried up in perfect safety to the height of 180 feet; and it has been used on a very large scale at the New Houses of Parliament now in progress.

Although of modern introduction in England this kind of scaffolding is not a new invention. It appears to have been used at Cologne Cathedral from the first com-mencement of that building in A.D 1248. It was also used by Domenic Fontana in A.D. 1586, for erecting the Egyptian Obelisk in front of St. Peter's at Rome; and similar scaffolding was used in Paris in our own times, in erecting the Arc de l'Etoile, and the Eglise de la Madcleine.

201 The moveable derrick crane is also much used in setting mason's work. It consists of a vertical post, supported by two timber backstays, and a long moveable jib or derrick hinged against the post below the gearing.

By means of a chain passing from a barrel over a pulley at the top of the post, the derrick can be raised to an almost vertical, or lowered to an almost horizontal position, thus enabling it to command every part of the area of a circle of a radius nearly equal to the length of the derrick. This gives it a great advantage over the old gibbet crane, which only commands a circle of a fixed radius, and the use of which entails great loss of time from its constantly requiring to be shifted as the work proceeds.

Derrick cranes appear to have been first introduced at Glasgow, A.D. 1833, by Mr. York, since which their original construction has been very greatly improved upon, and they are now very extensively used.

202. In hoisting blocks of stone they are attached to the tackle by means of a simple contrivance called a *lewis*, which is shown in fig. 84.

A tapering hole having been cut in the upper surface of the stone to be raised, the two side pieces of the lewis are inserted and placed against the sides of the hole;

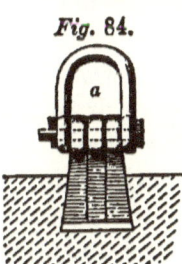

Fig. 84.

the centre parallel piece *a* is then inserted and secured in its place by a pin passing through all three pieces, and the stone may then be safely hoisted, as it is impossible for the lewis to draw out of the hole. By means of the lewis, in a slightly altered form from that here shown, stones can be lowered and set under water without difficulty, the lewis being disengaged by means of a line attached to the parallel piece; the removal of which allows the others to be drawn out of the mortice.

203 In stone-cutting, the workman forms as many plane

faces as may be necessary for bringing the stone into the required shape, with the least waste of material and labour, and on the plane surfaces so formed applies the moulds to which the stone is to be worked

To form a plane surface, the mason first knocks off the superfluous stone along one edge of the block, as

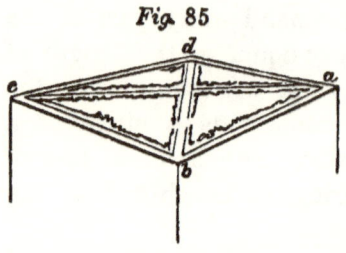

Fig. 85

a, b (fig. 85), until it coincides with a straight edge throughout its whole length; this is called a *chisel draught* Another chisel draught is then made along one of the adjacent edges, as *b, c,* and the ends of the two are connected by another draught, as *a, c*; a fourth draught is then sunk across the last, as *b, d,* which gives another angle point *d,* in the same plane with *a, b,* and *c,* by which the draughts *d a* and *a c* can be formed; and the stone is then knocked off between the outside draughts until a straight edge coincides with its surface in every part.

To form cylindrical or moulded surfaces curved in one

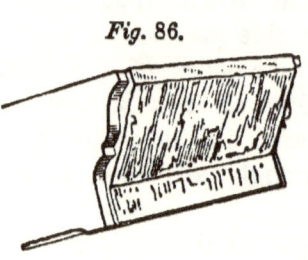

Fig. 86.

direction only, the workman sinks two parallel draughts at the opposite end of the stone to be worked, until they coincide with a mould cut to the required shape, and afterwards works off the stone between these draughts, by a straight edge applied at right angles to them (fig. 86).

The formation of conical or spherical surfaces is much less simple, and requires a knowledge of the scientific operations of stone-cutting, a description of which would be unsuited to the elementary character of these pages. The reader who wishes to pursue the subject is therefore referred to the volume of this series on "Masonry and Stone-cutting," where he will find the required information

204. The finely-grained stones are usually brought to a smooth face, and rubbed with sand to produce a perfectly even surface.

In working soft stones, the surface is brought to a smooth face with the *drag*, which is a plate of steel, indented on the edge like the teeth of a saw, to take off the marks of the tools employed in shaping it.

The harder and more coarsely-grained stones are generally *tooled*, that is, the marks of the chisel are left on their face. If the furrows left by the chisel are disposed in regular order, the work is said to be *fair-tooled*, but if otherwise, it may be *random-tooled*, or *chiselled*, or *boasted*, or *pointed*. If the stones project beyond the joints, the work is said to be *rusticated*.

Granite and gritstone are chiefly worked with the scappling hammer. In massive erections, where the stones are large, and a bold effect is required, the fronts of the blocks are left quite rough, as they come out of the quarry, and the work is then said to be *quarry pitched*.

Many technical terms are used by quarrymen and others engaged in working stone; but they need not be inserted here, as they are mostly confined to particular localities, beyond which they are little known, or perhaps bear a different signification.

205. When the mason requires to give to the joints of his work greater security than is afforded by the weight of the stone and the adhesion of the mortar, he makes use of *joggles*, *dowels*, and *cramps*.

Stones are said to be joggled together when a projection is worked out on one stone to fit into a corresponding hole or groove in the other (*see* fig. 87). But this occasions great labour and waste of stone, and *dowel-joggles* are chiefly made use of, which are hard pieces of stone, cut to the required size, and let into corresponding mortices in the two stones to bo joined together.

Fig. 87.

Dowels are pins of wood or metal used to secure the

joints of stone-work in exposed situations, as copings, pinnacles, &c. The best material is copper, but the expense of this metal causes it to be seldom used. If iron be made use of, it should be thoroughly tinned to prevent oxidation, or it will, sooner or later, burst and split the work it is intended to protect

Dowels are often secured in their places with lead poured in from above, through a small channel cut in the side of the joint for that purpose; but a good workman will eschew lead, which too often finds its way into bad work, and will prefer trusting to very close and workmanlike joints, carefully fitted dowels, and fine mortar; dowels should be made tapering at one end, which ensures a better fit, and renders the setting of the stone more easy for the workman.

Iron cramps are used as fastenings on the tops of copings, and in similar situations; but they are not to be recommended, as they are very unsightly, and, if they once become exposed to the action of the stmosphere, are powerfully destructive agents. Cast iron is, however, less objectionable than wrought iron for this purpose.

206. In measuring mason's work, the cubic content of the stone is taken as it comes to the *banker*, without deduction for subsequent waste.

If the scantlings are large, an extra price is allowed for hoisting.

The labour in working the stone is charged by the superficial foot, according to the kind of work, as plain work, sunk work, moulded work, &c

Pavings, landings, &c., and all stone less than 3 in. thick, are charged by the superficial foot.

Copings, curbs, window sills, &c., are charged per lineal foot.

Cramps, dowels, mortice holes, &c., are always charged separately.

A journeyman mason will receive from 4s. to 5s. 6d. per day, and the labourer from 2s. 6d. to 3s per day; but masons working at piece-work, or at any work requiring particular skill, will often earn much more.

The remuneration of a stone-carver is dependent on his talent, and the kind of work he is engaged upon

The following table of the weights of different kinds of stone will convey an idea of their relative hardness, and of the labour required to work them.

Table of the Weights of different kinds of Stone

13	cubic feet of marble	.	weigh one ton
13½	„	granite	„
14	„	Purbeck stone	„
14½	„	Yorkshire stone	„
16	„	Derbyshire grit	„
17	„	Portland stone	„
18	„	Bath stone	„

Mem.—58 ft. superficial of 3-in. York paving weigh one ton

70 ft. superficial of 2½-in. York paving weigh one ton

CARPENTER

207. The business of the carpenter consists in framing timbers together, for the construction of roofs, partitions, floors, &c.

208. The carpenter's principal tools are the axe, the adze, the saw, and the chisel, to which may be added the chalk-line, plumb-rule, level, and square. The work of the carpenter does not require the use of the plane, which is one of the principal tools of the joiner, and this forms the principal distinction beween these trades, the carpenter being engaged in the rough framework, and the joiner on the finishings and decorations of buildings.

209. The principles of framing have been already fully described in the 1st section of this work, and we shall therefore confine our remarks on the operations of the carpenter to a description of the principal joints made use of in framing.

Timbers that have to be joined in the direction of their length are *scarfed*, as shown in fig. 88; the double wedges, *a a*, serve to bring the timbers *home*, when they are secured,

Fig. 88.

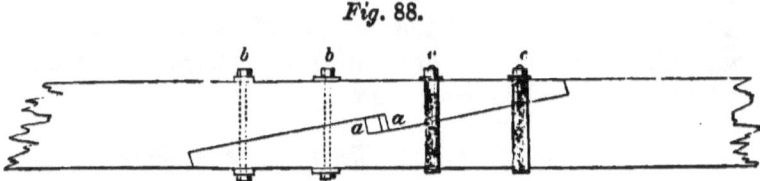

either by bolts, as shown at *b b*, or by straps, as at *c c*, the latter being the most perfect and the most expensive fastening.

Fig. 89 shows the manner of connecting the foot of a principal rafter with a tie-beam   The bolt here shown

Fig. 89.

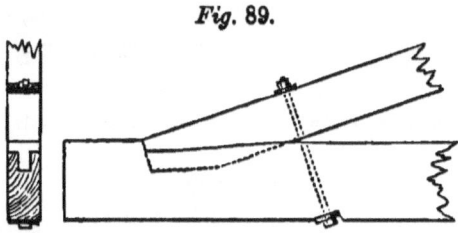

keeps the rafter in its place, and prevents it from slipping away from the abutment cut for it, which, by throwing the thrust on the tenon, would probably split it.   The end of the rafter should be cut with a square butt, so that the shrinkage of the timber will not lead to any settlement.

Fig. 90.

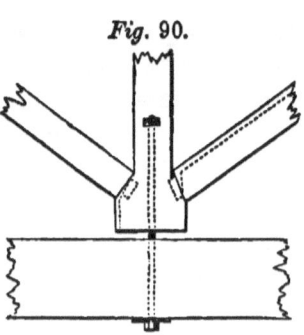

The connection of the foot of a king-post with the tie beam to be suspended from it is shown in fig. 90.

The king-post should be cut somewhat short, to give the power of screwing up the framing after the timber has become fully seasoned. The tie-beam may be suspended from the king-post, either by a bolt, as shown, or by a strap passed round the tie-beam and secured by iron wedges or cotters, passing through a hole in the king-post; this last is the more perfect, but at the same time the more expen sive of the two methods.

Fig. 90 also shows the manner in which the feet of the struts butt upon the king-post. They are slightly tenoned to keep them in their places. The ends of a strut should be cut ʃ as nearly square as possible, otherwise, when the timber shrinks, which it will always do, more or less, the thrust is thrown upon the edge only, which splits or crushes under the pressure, and causes settlement.

This is shown out by the dotted lines on the right-hand side of the cut. The dotted lines on the opposite side of the figure show a similar effect, produced by the shrinking of the king-post, for which there is no preventative but making it of oak, or some other hard wood. The same observations apply to the connections of the principal rafters with the top of the king-post, which are managed in a precisely similar manner.

In figures 91, 92, and 93, are shown different methods

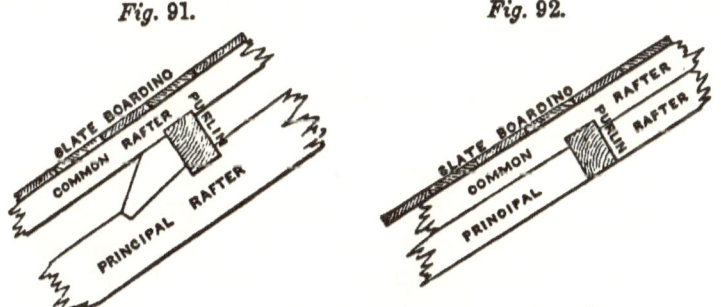

Fig. 91.   *Fig.* 92.

of fixing purlins, which are sufficiently explained by the figures to need no further description.

In figures 46, 47, 48, and 49, are shown the modes of framing the ends of binding joists into girders, and of connecting the ceiling joists with the binders; and as these have been already described under the head of "Floors,"

Fig. 93.

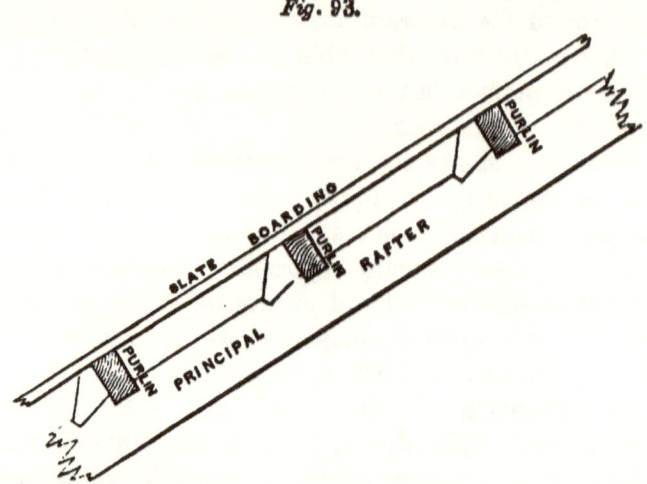

it is unnecessary here to say anything further on the subject.

As a general rule, all timbers should be notched down to those on which they rest, so as to prevent their being moved either lengthways or sideways. Where an upright post has to be fixed between two horizontal sills, as in the case of the uprights of a common framed partition, it is simply tenoned into them, and the tenons secured with oak pins driven through the cheeks of the mortice.

210. The carpenter requires considerable bodily strength for the handling of the timbers on which he has to work; he should have a knowledge of mechanics, that he may understand the nature of the strains and thrusts to which his work is exposed, and the best method of preventing or resisting them; and he should have such a knowledge of working drawings as will enable him, from the sketches of the architect, to set out the *lines* for every description of centering and framing that may be entrusted to him for execution.

211. In measuring carpenters' work the tenons are included in the length of the timber; this is not the case

in joiners' work, in which they are allowed for in the price

The labour in framing, roofs, partitions, floors, &c., is either valued at per square of 100 superficial feet, and the timber charged for separately, or the timber is charged as " fixed in place," the price varying according to the labour on it, as " cube fir in bond," " cube fir framed," " cube fir wrought and framed," &c. For shoring ⅛ of the value of the timber is allowed for use and waste.

The wages of a journeyman carpenter are from 4s. to 5s. 6d. per day.

JOINER.

212. The work of the joiner consists in framing and *joining* together the wooden finishings and decorations of buildings, both internal and external, such as floors, stair cases, framed-partitions, skirtings, solid door and window frames, hollow or *cased* window frames, sashes and shutters, doors, columns and entablatures, chimney-pieces, &c., &c.

The joiner's work requires much greater accuracy and finish than that of the carpenter, and differs materially from it in being brought to a smooth surface with the plane wherever exposed to view, whilst in carpenters' work the timber is left rough as it comes from the saw.

213. The joiner uses a great variety of tools; the prin-cipal *cutting* tools are *saws*, *planes*, and *chisels*.

Of saws there are many varieties, distinguished from each other by their shape and by the size of the teeth.

The *ripper* has 8 teeth in 3 inches; the *half-ripper* 3 teeth to the inch; the *hand saw* 15 teeth in 4 inches; the *panel saw* 6 teeth to the inch.

The *tenon saw*, used for cutting tenons, has about 8 teeth to the inch, and is strengthened at the back by a thick piece of iron, to keep the blade from buckling. The *sash saw* is similar to the tenon saw, but is backed with brass instead of iron, and has 13 teeth to the inch. The *dovetail saw* is still smaller, and has 15 teeth to the inch.

Besides the above, other saws are used for particular purposes, as the *compass saw*, for cutting circular work and the *key-hole saw*, for cutting out small holes. The *carcase saw* is a large kind of dovetail saw, having about 11 teeth to an inch.

214. Planes are also of many kinds; those called *bench planes*—as the *jack plane*, the *trying plane*, the *long plane*, the *jointer*, and the *smoothing plane*, are used for bringing the stuff to a plane surface. The jack plane is about 18 in. long, and is used for the roughest work. The trying plane is about 22 in. long, and used after the jack plane for *trying up*, that is, taking off shavings the whole length of the stuff; whilst in using the jack plane the workman stops at every arm's-length. The *long plane* is 2 ft. 3 in. long, and is used when a piece of stuff is to be tried up very straight. The *jointer* is 2 ft. 6 in. long, and is used for trying up or *shooting* the *joints*, in the same way as the trying plane is used for trying up the *face* of the stuff. The *smoothing plane* is small, being only 7½ in. long, and is used on almost all occasions for cleaning off finished work

Rebate planes are used for sinking *rebates* (*see* fig. 94), and vary in their size and shape according to their respec-
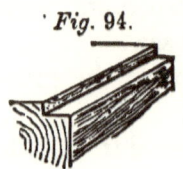
Fig. 94.
tive uses. Rebate planes differ from bench planes in having no handle rising out of the stock, and in discharging their shavings at the side. Amongst the rebate planes may be mentioned the *moving fillister* and the *sash fillister*, the uses of which will be better understood by inspection than from any description.

Moulding planes are used for *sticking* mouldings, as the operation of forming mouldings with the plane is called. When mouldings are worked out with chisels instead of with planes, they are said to be worked *by hand*. Of the class of moulding planes, although kept separate in the tool chest, are *hollows* and *rounds* of various sizes.

There are other kinds of planes besides the above; as the *plough*, for sinking a groove to receive a projecting tongue; the *bead plane*, for sticking beads; the *snipe bill*, for forming quirks; the *compass plane* and the *forkstaff plane*, for forming concave and convex cylindrical surfaces. The shape and use of these and many other tools used by the joiner will be better understood by a visit to the joiner s shop than by any verbal description.

215. Chisels are also varied in their form and use. Some are used merely with the pressure of the hand, as the *paring chisel;* others, by the aid of the mallet, as the *socket chisel,* * for cutting away superfluous stuff; and the *mortice chisel*, for cutting mortices. The *gouge* is a curved chisel.

216. The joiner uses a great variety of boring tools, as the *brad-awl, gimlet,* and *stock and bit.* The last form but one tool, the *stock* being the handle, to the bottom of which may be fitted a variety of steel bits of different bores and shapes, for boring and widening out holes in wood and metal, as *countersinks, rimers,* and *taper shell bits.*

217. The *screw-driver, pincers, hammer, mallet, hatchet,* and *adze,* are too well known to need description.

The *gauge* is used for drawing lines on a piece of stuff parallel to one of its edges.

The *bench* is one of the most important of the joiner's implements. It is furnished with a vertical *sideboard,* perforated with diagonal ranges of holes, which receive the *bench pin* on which to rest the lower end of a piece of stuff to be planed, whilst the upper end is firmly clamped by the *bench screw.*

The *mitre box* is used for cutting a piece of stuff to a *mitre* or angle of 45 degrees with one of its sides.

The joiner uses for setting out and fixing his work—the straight edge, the square, the bevel or square with a shifting blade, the mitre square, the level, and the plumb rule.

In addition to the tools and implements above enumerated, the execution of particular kinds of work requires

* Named from the iron forming a socket to receive a wooden handle.

other articles, as cylinders, templets, cramps, &c., the
description of which would unnecessarily
extend the limits of this volume

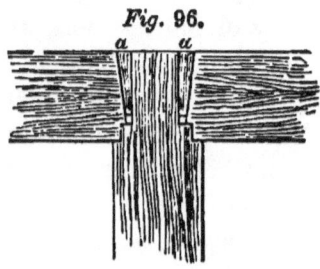

Fig. 95.

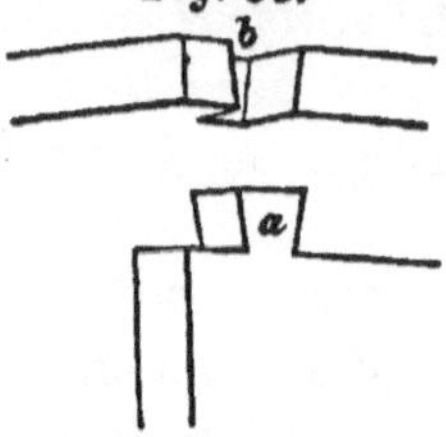

218. The principal operations of the
joiner are sawing, planing, dovetailing,
mortising, and scribing.

The manner of forming a *dovetail* is
shown in fig. 95. The projecting part,
a, is called the *pin*, and the hole to
receive it is called the *socket*.

Mortising is shown in fig. 96; the projecting piece is
called the *tenon*, and the hole
formed to receive it the *mortice*.
The tenon is sometimes *pinned*
in its place with oak pins driven
through the cheeks of the mortice, but in forming doors, shutters, &c., the tenon is secured
with tapering wedges driven into
the mortice, which is cut slightly
wider at the top than at the bottom, the adhesion of the
glue with which the wedges are first rubbed over, making
it impossible for the tenon afterwards to draw out of its
place.

Fig. 96.

219. Joints in the length of the stuff may be either
square, as at *a*, fig. 97, or rebated, as at *b*, or grooved and

Fig. 97.

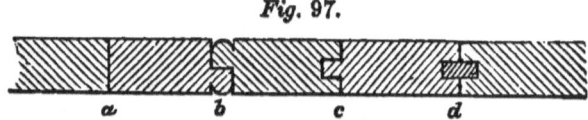

tongued, as at *c*, or grooved on each edge and a tongue let
in, as at *d*

220. *Scribing* is the drawing on a piece of stuff the exact
profile of some irregular surface to which it is to be made
to fit: this is done with a pair of compasses, one leg of
which is made to traverse the irregular surface, the other to

describe a line parallel thereto along the edge of the stuff to be cut.

221. In the execution of circular, or, as it is termed, *sweep work*, there are four different methods by which the stuff can be brought to the required curve :—

1st. It may be steamed and bent into shape.

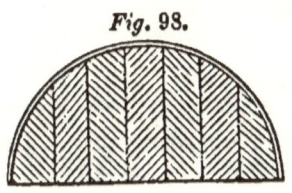

Fig. 98.

2nd. It may be glued up in thicknesses, as shown in fig. 98, which must, when thoroughly dry, be planed true, and, if not to be painted, covered with a thin veneer bent round it.

3rd. It may be formed in thin thicknesses, as shown in fig 99, bent round and glued up in a mould. This may be considered the most perfect of all the methods in use

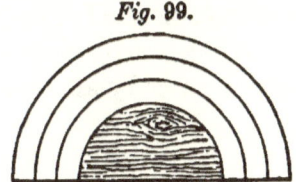

Fig. 99.

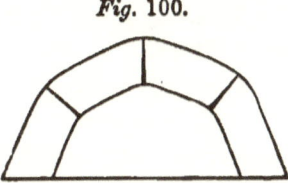

Fig. 100.

Lastly It may be formed by sawing a number of notches on one side, as shown in fig. 100, by which means it becomes easily bent in that direction, but the curve produced by this means is very irregular, and it is an inferior mode of execution compared to the others.

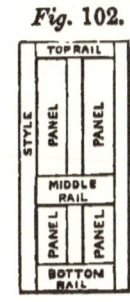

Fig. 101. *Fig. 102.*

222. When a number of boards are secured together by cross-pieces or *ledges* nailed or screwed at the back, the work is said to be *ledged* (*see* fig. 101). Ledged work is used for common purposes, as cellar doors, outside shutters, &c.

Framed work (fig. 102) consists of *styles* and *rails* mortised and tenoned together, and filled in with pannels, the edges of which fit in grooves cut for that purpose in the styles and rails.

Work is said to be *clamped* when it is prevented from warping or splitting by a rail at each

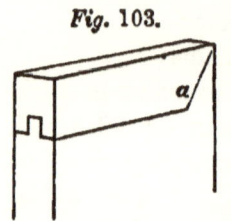

Fig. 103.

end, as in fig. 103; if the ends of the rail are cut off, as shown at *a*, it is said to be mitre clamped.

223. There are two ways of laying floors practised by joiners. In laying what is called a *straight joint* floor, from the joints between the boards running in an unbroken line from wall to wall, each board is laid down and nailed in succession, being first forced firmly against the one last laid with a flooring cramp.

Folding floors are laid by nailing down first every fifth board rather closer together than the united widths of four boards, and forcing the intermediate ones into the space left for them by jumping upon them; this method of laying floors is resorted to when the stuff is imperfectly seasoned and is expected to shrink, but it should never be allowed in good work.

The narrower the stuff with which a floor is laid the less will the joints open, on account of the shrinkage being distributed over a greater number of joints.

The floor boards may be nailed at their edges, and grooved and tongued or dowelled, if it be wished to make a very perfect floor. Dowelling is superior to grooving and tonguing, because the cutting away the stuff to receive the tongue greatly weakens the edges of the joint, which are apt to curl

224. Glue is an article of great importance to the joiner; the strength of his work depending much upon its adhesive properties.

The best glue is made from the *skins* of animals; that from the *sinewy* or *horny* parts being of inferior quality. The strength of the glue increases with the age of the animals from which the skins are taken.

225. Joiners' work is measured by the superficial foot, according to its description.

Floors by the square of 100 superficial feet.

Handrails, small mouldings, water-trunks, and similar articles, per lineal foot.

Cantilevers, trusses, cut brackets, scrolls to handrails, &c., are valued per piece.

The wages of a joiner are from 4s. to 5s. 6d. per day.

The following memoranda relative to carpenters' and joiners' work may be found useful

Weight of Timber

34 cubic feet of mahogany weigh one ton

39	„	oak	„	„
45	„	ash	„	„
51	„	beech	„	„
60	„	elm	„	„
65	„	fir	„	„

50 cubic feet of timber 1 load.

120 deals = one hundred.

120 12 ft. 3 in. deals = $5\frac{2}{5}$ loads of timber.

400 superficial feet $1\frac{1}{2}$ in. deal = 1 load.

Planks are . . . . 11 in. wide

Deals . . . . 9 „

Battens . . . . 7 „

A reduced deal is $1\frac{1}{2}$ in. thick, 11 in wide, and 12 feet long.

A square of flooring laid with 12 feet deals requires

Laid rough . . . . . . $12\frac{1}{4}$ floorboards.

Ditto, edges shot . . . . . $12\frac{1}{2}$ „

Wrought and laid folding . . . 13 „

Ditto, straight joint . . . . $13\frac{1}{2}$ „

Wrought and laid straight joint, and

ploughed and tongued . . . 14 „

If laid with 12 ft. battens,

Wrought, and laid folding . . 17 „

Ditto, ditto, straight joint . . 18 „

226. *Ironmongery* is charged for with the work to which it is attached; the joiner being allowed 20 per cent. profit upon the prime cost.

The principal articles of ironmongery used in a building consist of *nails* and *screws, sash pullies, bolts, hinges, locks, latches,* and *sash* and *shutter furniture,* besides a great variety of miscellaneous articles, which we have not space to enumerate.

227. Of the different kinds of hinges may be mentioned *hook* and *eye hinges,* for gates, coach-house doors, &c.; *butts* and *back-flaps,* for doors and shutters; *cross garnets* of ⊢ form, which are used for hanging ledged doors, and other inferior work: H and H— hinges, whose name is derived from their shape; and *parliament* hinges.

Besides these are used *rising butts,* for hanging doors to rise over a carpet, or other impediment; *projecting butts,* used when some projection has to be cleared, and *spring hinges* and *swing centres,* for self-shutting doors.

228. The variety of locks now manufactured is almost infinite. We may mention the *stock lock,* cased in wood, for common work. *Rim locks,* which have a metal case or rim, and are attached to one side of a door: they should not be used when a door has sufficient thickness to allow of a mortice lock, as they often catch the dresses of persons passing through the doorway. *Mortice locks,* as the name implies, are those which are mortised into the thickness of the door.

The handles and escutcheons are called the *furniture* of a lock, and are made of a great variety of materials, as brass, bronze, ebony, ivory, glass, &c.

229. Of latches, there are the common *thumb latch,* the *bow latch,* with brass knobs, the brass *pulpit latch,* and the *mortice latch.*

230. The *sawyer* is to the carpenter and joiner what the stone-cutter is to the mason

The *pit-saw* is a large two-handed saw fixed in a frame, and moved up and down in a vertical direction, by two men called the top-man and the pit-man; the first of whom stands on the timber that is to be cut, the other at the bottom of the saw-pit. The timber is *lined out* with a chalk line on its upper surface, and the accuracy of the work depends mainly on the top-man keeping the saw to the line, whence the proverbial expression *top-sawyer*, meaning one who directs any undertaking.

In sawing up deals and battens into thicknesses for the joiner's use, the parallelism of the cuts is of the utmost importance, as the operation of *taking out of winding*, a piece of uneven stuff, causes a considerable waste of material, and much loss of time.

Circular saws, moved by steam power,* are now much used in large establishments, timber yards, &c., and effect a great saving of labour over the use of the pit saw, where the timbers to be cut are not too heavy to be easily handled. The saw is mounted in the middle of a stout bench, furnished with guides, by means of which the stuff to be cut is kept in the required direction, whilst it is pushed against the saw, which is the whole of the manual labour required in the operation.

SLATER.

231. The business of the slater consists chiefly in covering the roofs of houses with slates, but it has of late years been very much extended by the general introduction of sawn slate, as a material for shelves, cisterns, baths, chimney pieces, and even for ornamental purposes.

We purpose here to describe only those operations of the slater which have reference to the covering of roofs.

232. Besides the tools which are in common use among other artificers, the slater uses one peculiar to his trade, called the *zax*, which is a kind of hatchet, with a sharp point

---

* The author recently visited a carpenter's shop in a country village in Leicestershire, which was mounted in a very complete manner, with bench and other saws, lathes, &c., all worked by a set of wind-sails on the roof.

at the back. It is used for trimming slates, and making the holes by which they are nailed in their places.

233. Slates are laid either on boarding or on narrow battens, from 2 to 3 inches wide, the latter being the more common method, on account of its being less expensive than the other.

The nails used should be either copper or zinc; iron nails, though sometimes used, being objectionable from their liability to rust.

Every slate should be fastened with two nails, except in the most inferior work.

The upper surface of a slate is called its *back*, the under surface the *bed*, the lower edge the *tail*, the upper edge the *head*. The part of each course of slates exposed to view is called the *margin* of the course, and the width of the margin is called the *gauge*.

The *bond* or *lap* is the distance which the lower edge of any course overlaps the slates of the second course below, measuring from the nail-hole.

In preparing slates for use, the sides and bottom edges are trimmed, and the nail-holes punched as near the head as can be done, without risk of breaking the slate, and at a uniform distance from the tail.

The lap having been decided on, the gauge will be equal to half the distance from the tail to the nail-hole, less the lap. Thus a countess slate, measuring 19 in. from tail to nail, if laid with a 3-in. lap, would show a margin of

$$-\frac{19 \text{ in.} - 3 \text{ in.}}{2} = 8 \text{ in. } (See \text{ figs. } 104, 105.)$$

Fig. 104.   Fig. 105.

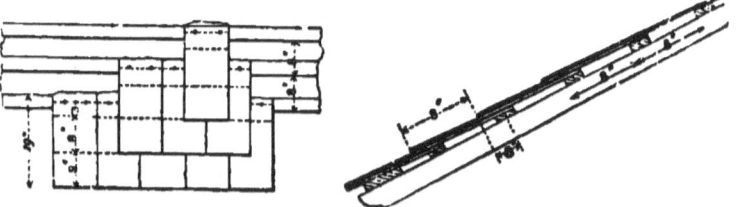

The battens are of course nailed on the rafters at the gauge to which the slates will work. If the slates are of different lengths, they must be sorted into sizes, and gauged accordingly, the smallest sizes being placed nearest the ridge. The lap should not be less than 2 in., and need not exceed 3 in

It is essential to the soundness as well as the appearance of slaters' work, that the slates should all be of the same width, and the edges perfectly true.

The Welsh slates are considered the best, and are of a light sky blue colour. The Westmoreland slates are of a dull greenish hue.

234. Slaters' work is measured by the square of 100 superficial feet, allowances being made for the trouble of cutting the slates at the hips, eaves, round chimneys, &c.

Slabs for cisterns, baths, shelves, and other sawn work, are charged per superficial foot, according to the thickness of the slab, and the labour bestowed on the work.

Rubbed edges, grooves, &c., are charged per lineal foot.

Table of the Sizes of Roofing Slates.

DESCRIPTION.	Size.		Average gauge in inches.	No. of squares 1200 will cover.	Weight per 1200 in tons.	No. required to cover one square.	No. of nails required to one square.
	Length.	Breadth.					
	ft. in.	ft. in.					
Doubles . . .	1 1	0 6	5½	2	¾	480	480
Ladies . . .	1 4	0 8	7	4½	1¼	280	280
Countesses . .	1 8	0 10	9	7	2	176	352
Duchesses . .	2 0	1 0	10½	10	3	127	254
Imperials . .	2 6	2 0	} a ton will cover 2¼ to 2½ squares.				
Rags and Queens	3 0	2 0					
Westmorelands, of various sizes			do.	do.	2	do.	

Inch slab per foot superficial weighs 14 lbs.

A journeyman slater receives about 5s. per day, and hi labourer about 3s.

PLASTERER

285. The work of the plasterer consists in covering the brickwork and naked timbers of walls, ceilings, and partitions with plaster, to prepare them for painting, papering, or distempering; and in forming cornices, and such decorative portions of the finishings of buildings as may be required to be executed in plaster or cement.

236. The plasterer uses a variety of tools, of which the following are the principal ones:—

The *drag* is a three-pronged rake, used to mix the hair with the mortar in preparing coarse stuff.

The *hawk* is a small square board for holding stuff on, with a short handle on the under side.

Trowels are of two kinds, the *laying and smoothing tool*, with which the first and the last coats are laid, and the *gauging trowel*, used for gauging fine stuff for cornices, &c.; these are made of various sizes, from 3 to 7 in. long.

Of *floats*, which are used in *floating*, there are three kinds, viz. the *Derby*, which is a rule of such a length as to require two men to use it; the *hand float*, which is used in finishing stucco; and the *quirk float*, which is used in floating angles.

Moulds, for running cornices, are made of sheet copper, cut to the profile of the moulding to be formed, and fixed in a wooden frame.

Stopping and *picking out tools* are made of steel, 7 or 8 in long, and of various sizes. They are used for modelling, and for finishing mitres and returns to cornices.

237. Materials.—*Coarse stuff*, or lime and hair, as it is usually called, is similar to common mortar, with the addition of hair from the tanner's yard, which is thoroughly mixed with the mortar by means of the drag.

Fine stuff is made of pure lime, slaked with a small quantity of water, after which, sufficient water is added to bring it to the consistence of cream.

It is then allowed to settle, and the superfluous water being poured off, it is left in a binn or tub to remain in a semifluid state until the evaporation of the water has brought it to a proper thickness for use. In using fine stuff for setting ceilings, a small portion of white hair is mixed with it.

Stucco is made with fine stuff, and clean-washed sand This is used for finishing work intended to be painted.

Gauged stuff is formed of fine stuff mixed with plaster of Paris, the proportion of plaster varying according to the rapidity with which the work is required to set. Gauged stuff is used for running cornices and mouldings.

Enrichments, such as pateras, centre flowers for ceilings &c., are first modelled in clay, and afterwards cast of plaster of Paris in wax or plaster moulds. Papier maché ornaments also are much used, and have the advantage of being very light, and being easily and securely fixed with screws.

The variety of compositions and cements made use of by the plasterer is very great. Roman cement, Portland cement, and lias cement, are the principal ones used for coating buildings externally. Martin's and Keene's cements are well adapted for all internal plastering where sharpness, hardness, and delicate finish are required.

238. *Operations of Plastering.*—When brickwork is plastered, the first coat is called *rendering*.

In plastering ceilings and partitions, the first operation is *lathing*. This is done with *single, one and a half*, or *double* laths; these names denoting their respective thicknesses Laths are either of oak or fir; if the former, wrought-iron nails are used, but cast-iron nails may be employed with the latter. The thickest laths are used for ceilings, as the strain on the laths is greater in a horizontal than in an upright position.

Pricking up is the first coat of plastering of coarse stuff upon laths; when completed, it is well scratched over with the end of a lath, to form a key for the next coat.

Laid work consists of a simple coat of coarse stuff over a wall or ceiling.

Two-coat work is a cheap description of plastering, in which the first coat is only roughed over with a broom, and afterwards *set* with fine stuff, or with gauged stuff in the better descriptions of work.

The laying on of the second coat of plastering is called *floating*, from its being *floated*, or brought to a plane surface with the float.

The operation of floating is performed by surrounding the surface to be floated with narrow strips of plastering, called screeds, brought perfectly upright, or level, as the case may be, with the level or plumb-rule; thus, in preparing for floating a ceiling, nails are driven in at the angles, and along the sides, about 10 ft. apart, and carefully adjusted to a horizontal plane, by means of the level. Other nails are then adjusted exactly opposite to the first, at a distance of 7 or 8 in. from them. The space between each pair of nails is filled up with coarse stuff, and levelled with a hand float; this operation forms what are called *dots* When the dots are sufficiently dry, the spaces between the dots are filled up flush with coarse stuff, and floated perfectly true with a floating rule; this operation forms a *screed*, and is continued until the ceiling is surrounded by one continuous screed, perfectly level throughout. Other screeds are then formed, to divide the work into bays about 8 ft. wide, which are successively filled up flush, and floated level with the screeds.

The screeds for floating walls are formed in exactly the same manner, except that they are adjusted with the plumb-rule instead of the level.

After the work has been brought to an even surface with the floating rule, it is gone over with the hand float, and a little soft stuff, to make good any deficiencies that may appear

The operation of forming screeds and floating work, which is not either vertical or horizontal, as a plaster floor laid with a fall, is analogous to that of taking the face of a stone out of winding with chisel-drafts and straight edges in stone-cutting; the principle being in each case to find three points in the same plane from which to extend operations over the whole surface

Setting —When the floating is about half dry, the setting or finishing coat of fine stuff is laid on with the smoothing trowel, which is alternately wetted with a brush and worked over with the smoothing tool, until a fine surface is ob tained.

Stucco is laid on with the largest trowel, and worked over with the hand float, the work being alternately sprinkled with water, and floated until it becomes hard and compact, after which it is finished by rubbing it over with a dry stock brush.

The water has the effect of hardening the face of the stucco, so that, after repeated sprinklings and trowelings, it becomes very hard, and smooth as glass.

239. The above remarks may be briefly summed up as follows. The commonest kind of work consists of only one coat, and is called *rendering*, on brickwork, and *laying*, if on laths. If a second coat be added, it becomes two-coat work, as *render-set*, or *lath lay* and *set*. When the work is floated, it becomes three-coat work, and is *render, float*, and *set*, for brickwork, and *lath, lay, float*, and *set*, for ceilings and partitions ; ceilings being set with fine stuff, with a little white hair, and walls intended for paper with fine stuff and sand ; stucco is used where the work is to be painted.

Rough stucco is a mode of finishing staircases, passages, &c., in imitation of stone. It is mixed with a large proportion of sand, and that of a coarser quality than troweled stucco, and is not smoothed, but left rough from the hand float, which is covered with a piece of felt, to raise the grit of the sand, to give the work the appearance of stone.

Rough cast is a mode of finishing outside work, by dashing over the second coat of plastering, whilst quite wet, a layer of rough-cast, composed of well-washed gravel, mixed up with pure lime and water, till the whole is in a semi-fluid state.

Pugging is lining the spaces between floor joists with coarse stuff, to prevent the passage of sound, or between two stones, and is done on laths or rough boarding.

In the midland districts of England, reeds are much used instead of laths, not only for ceilings and partitions, but for floors, which are formed with a thick layer of coarse gauged stuff upon reeds. Floors of this kind are extensively used about Nottingham; and, from the security against fire afforded by the absence of wooden floors, Nottingham houses are proverbially fire-proof

240. Plasterer's work is measured by the superficial yard; cornices by the superficial foot; enrichments to cornices by the lineal foot; and centre flowers and other decorations at per piece.

The wages of a journeyman plasterer are from 4s. to 5s a day; those engaged in modelling and ornamental work will earn much more; a labourer receives from 2s. 6d. to 3s a day, and a plasterer's boy about 1s.

Lathing.—One bundle of laths and 384 nails will cover 5 yards.

Rendering.—187½ yards require 1½ hundred of lime, 2 double loads of sand, and 5 bushels of hair.

Floating requires more labour, but only half as much material as rendering.

Setting.—375 yards require 1½ hundred of lime, and 5 bushels of hair.

Render set.—100 yards require 1½ hundred of lime, 1 double load of sand, and 4 bushels of hair.—Plasterer, labourer, and boy, three days each.

Lath, lay, and *set.*—130 yards of lath, lay, and set, require 1 load of laths, 10,000 nails, 2½ hundred of lime, 1¼ double load of sand, and 7 bushels of hair.—Plasterer, labourer, and boy, six days each.

Twenty per cent. profit is allowed on all materials.

SMITH AND IRONFOUNDER

241. The smith furnishes the various articles of wrought-iron work used in a building; as pileshoes, straps, screw-

bolts, dog-irons, chimney bars, gratings, wrought-iron rail ing, and wrought-iron balustrades for staircases. Wrought iron was formerly much used for many purposes for which cast iron is now almost exclusively employed; the improvements effected in casting during the present century having made a great alteration in this respect.

The operations of the ironfounder have been described in Section II. of this volume, and therefore we have only here to enumerate some of the principal articles which are furnished by him.

Besides cast-iron columns, girders, and similar articles which are cast to order, the founder supplies a great variety of articles which are kept in store for immediate use; as cast-iron gratings, balconies, rain-water pipes and guttering, air traps, coal plates, stoves, stable fittings, iron sashes, &c.

Both wrought and cast iron work are paid for by weight, except small articles kept in store for immediate use, which are valued per piece.

			lbs.
One cubic foot of cast iron weighs about			450
Ditto	wrought	„	475
Ditto	closely hammered		485

242. The *coppersmith* provides and lays sheet copper for covering roofs; copper gutters, and rain-water pipes; washing and brewing coppers; copper cramps and dowels for stonemasons' work; and all other copper work in a building; but the cost of the material in which he works prevents its general use; and the washing copper is frequently the only part of a building which requires the aid of this artificer. Sheet copper is paid for by the superficial foot, according to weight, and pipes and gutters per lineal foot; copper in dowels, bolts, &c., at per pound.

243. *Warming apparatus, steam* and *gas fittings*, and similar kinds of work, are put up by the mechanical engineer, who also manufactures a great variety of articles, which are purchased in parts, and put together and fixed by the plumber, as pumps, taps, water closet apparatus, &c

244. The *bell-hanger* provides and hangs the bells required for communicating between the different parts of a building, and connects them with their *pulls*, or handles, by means of cranks and wires.

The action of the pull upon the bell should be as direct, and effected with as few cranks, as possible; and the cranks and wires should be concealed from view, both to protect them from injury, and on account of their unsightly appearance

In all superior work, the wires are conducted along concealed tubes, fixed to the walls before the plasterer's work is commenced. The simplest way of arranging the wires is to carry them up in separate tubes to the roof, where they may all be conducted to one point, and brought down a chase in the walls to the part of the basement where the bells are hung. By this means very few cranks are required, and a broken wire can be replaced at any time without trouble.

245. Bell-hangers' work is paid for by the number of bells hung; the price being determined by the manner in which the work is executed. The *furniture* to the pulls is charged in addition, at per piece

A journeyman smith receives about 5s. a day, and his labourer about 3s. 6d.; a good bell-hanger will receive 7s. a day

PLUMBER.

246. The work of the plumber chiefly consists in laying sheet lead on roofs, lining cisterns, laying on water to the different parts of a building, and fixing up pumps and water closets.

247. The plumber uses but few tools, and those are of a simple character; the greater number of them being similar to those used by other artificers, as *hammers*, *mallets*, *planes*, *chisels*, *gouges*, *files*, &c. The principal tool peculiar to the trade of the plumber is the *bat*, which is made of beech, about 18 in. long, and is used for dressing and flattening sheet lead For soldering also the plumber uses iron

ladles, of various sizes, for melting solder, and *grozing irons*, for smoothing down the joints.

248. The sheet lead used by the plumber is either *cast* or *milled*, the former being generally cast by the plumber himself out of old lead taken in exchange; whilst the latter, which is cast lead, flattened out between rollers in a flatting mill, is purchased from the manufacturer. Sheet lead is described according to the weight per superficial foot, as 5-lb. lead, 6-lb. lead, &c.

Lead pipes, if of large diameter, are made of sheet lead, dressed round a wooden core, and soldered up.

Smaller pipes are cast in short lengths, of a thickness three or four times that of the intended pipe, and either *drawn* or *rolled* out to the proper thickness.

Soft solder is used for uniting the joints of lead-work. It is made of equal parts of lead and tin, and is purchased of the manufacturer by the plumber, at a price per lb., according to the state of the market.

249. *Laying of Sheet Lead.*—In order to secure lead-work from the injurious effects of contraction and expansion, when exposed to the heat of the sun, the plumber is careful not to confine the metal by soldered joints, or otherwise. All sheet lead should be laid to a sufficient *current*, to keep it dry; a fall of 1 in. in 10 ft. is sufficient for this purpose, if the boarding on which the lead is laid be perfectly even Joints in the direction of the current are made by dressing the edges of the lead over a wooden *roll*, as shown in fig. 106.

Joints in the length of the current are made with *drips*, as shown on the left-hand side of fig. 107

Fig. 106. *Fig.* 107.

Flashings are pieces of lead *turned down* over the edges of other lead work, which is *turned up* against a wall, as shown on the right-hand side of fig. 107, and serve to keep the wet from finding its way between the wall and the lead

The most secure way of fixing them is to build them into the joints of the brickwork; but the common method is to insert them about an inch into the mortar joint, and to secure them with wall hooks and cement. (*See* fig. 107.'

250. A very important part of the business of the plumber consists in fitting up cisterns, pumps, and water-closet apparatus, and in laying the different services and wastes connected with the same.

251. Plumbers' work is paid for by the cwt., milled lead being rather more expensive than cast.

Lead pipes are charged per foot lineal, according to size.

Pumps and water-closet apparatus are charged at so much each, according to description; as also basins, air traps, washers and plugs, spindle valves, stop-cocks, ball-cocks, &c.

Table of the Weight of Lead Pipes, per yard.

Bore.		lbs.	oz.
$\frac{1}{2}$ inch	. . . . . .	3	3
$\frac{3}{4}$,,	. . . . .	5	7
1 ,,	. . . . .	8	0
$1\frac{1}{4}$,,	. . . . .	11	0
$1\frac{1}{2}$,,	. . . .	14	0
2 ,,	. . . . .	21	0

The wages of a journeyman plumber are from 5s. to 6s. a day The plumber's labourer receives from 3s. to 3s. 6d. a day

ZINC WORKER.

252. The use of sheet lead has been, to a certain extent, superseded by the use of sheet zinc, which, from its cheapness and lightness, is very extensively used for almost all purposes to which sheet lead is applied. It is, however, a very inferior material, and not to be depended upon. The laying of it is generally executed by the plumber; but the working of zinc, and manufacturing of it into gutters, rainwater pipes, chimney cowls. and other articles, is practised as a distinct business

GLAZIER.

253. The business of the glazier consists in cutting glass, and fixing it into lead-work, or sashes. The former is the

oldest description of glazing, and is still used, not only for cottage windows, and inferior work, but for church windows, and glazing with stained glass, which is cut into pieces of the required size, and set in a leaden framework; this kind of glazing is called *fretwork*.

254. *Glazing in sashes* is of comparatively modern introduction. The sash-bars are formed with a *rebate* on the outside, for the reception of the glass, which is *cut into* the rebates, and firmly *bedded* and *backputtied* to keep it in its place. Large squares are also *sprigged*, or secured with small brads driven into the sash-bars.

255 *Glazing in lead-work* is fixed in leaden rods, called *cames*, prepared for the use of the glazier by being passed through a glazier's vice, in which they receive the grooves for the insertion of the glass. The sides or cheeks of the grooves are sufficiently soft to allow of their being turned down to admit the glass, and again raised up and firmly pressed against it after its insertion.

For common lead-work, the bars are soldered together, so as to form squares or diamonds. In fretwork, the bars, instead of being used straight, are bent round to the shapes of the different pieces of glass forming the device—lead-work is strengthened by being attached to *saddle bars* of iron, by leaden bands soldered to the lead-work, and twisted round the iron.

Putty is made of pounded whiting, beaten up with linseed oil into a tough tenacious cement.

256. The principal tool of the glazier is the *diamond*, which is used for cutting glass. This tool consists of an unpolished diamond fixed in lead, and fastened to a handle of hard wood.

The glazier uses a *hacking-out knife*, for cutting out old putty from broken squares; and the *stopping knife* for laying and smoothing the putty when *stopping-in* glass into sashes.

For setting glass into lead-work, the *setting knife* is used.

Besides the above, the glazier requires a square and

straight edges, a rule, and a pair of compasses, for dividing the tables of glass to the required sizes.

Also a hammer and brushes, for sprigging large squares, and cleaning off the work.

The *glazier's vice* has already been mentioned; the *latter-kin* is a pointed piece of hard wood, with which the grooves of the *cames* are cleared out and widened for receiving the glass.

257. Cleaning windows is an important branch of the glazier's business in most large towns; the glazier taking upon himself the cost of repairing all glass broken in cleaning.

258. Glaziers' work is valued by the superficial foot, the price increasing with the size of the squares. Irregular panes are taken of the extreme dimensions each way.

Crown glass is *blown* in circular *tables* from 3 ft. 6 in. to 5 ft. diameter, and is sold in *crates*, the number of tables in a crate varying according to the quality of the glass.

A crate contains 12 tables of best quality
„   „   15   „   second do.
„   „   18   „   third do.

Plate glass is *cast* in large plates on horizontal tables, and afterwards polished

The manufacture of sheet or spread glass, which was formerly considered a very inferior article, has of late years been much improved: much is now sold, after being polished, under the name of Patent Plate

PAINTER, PAPER-HANGER, AND DECORATOR.

259. The business of the house-painter consists in covering, with a preparation of white lead and oil, such portions of the joiner's, smith's, and plasterer's work as require to be protected from the action of the atmosphere. Decorative painting is a higher branch, requiring a knowledge of the harmony of colours, and more or less of artistic skill, according to the nature of the work to be executed. The introduction of fresco painting into this country as a mode

of internal decoration has led to the employment of some of the first artists of the day in the embellishment of the mansions of the nobility; and the example thus set will, no doubt, be extensively followed.

260. The principal materials used by the painter arc *white lead*, which forms the basis of almost all the colours used in house-painting; *linseed oil*, and *spirits of turpentine*, used for mixing and diluting the colours; and *dryers*, as litharge, sugar of lead, and white vitriol, which are mixed with the colours to facilitate their drying *Putty*, made of whiting and linseed oil, is used for *stopping* or filling up nail holes, and other vacuities, in order to bring the work to a smooth face.

261. The painter's tools are few and simple; they consist of the *grinding stone* and *muller*, for grinding colours; *earthen pots*, to hold colours; *cans*, for oil and turps; a *pallet knife*, and *brushes* of various sizes and descriptions.

262. In painting woodwork, the first operation consists in *killing* the knots, from which the turpentine would otherwise exude and spoil the work. To effect this, the knots are covered with fresh slaked lime, which dries up and burns out the turpentine. When this has been on twenty-four hours, it is scraped off, and the knots painted over with a mixture of red and white lead, mixed with glue size. After this they are gone over a second time with red and white lead, mixed with linseed oil. When dry, they must be rubbed perfectly smooth with pumice stone, and the work is ready to receive the priming coat. This is composed of red and white lead, well diluted with linseed oil. The nail holes and other imperfections are then stopped with putty, and the succeeding coats are laid on, the work being rubbed down between each coat, to bring it to an even surface. The first coat after the priming is mixed with linseed oil and a little turpentine; the second coat with equal quantities of linseed oil and turpentine. In laying on the second coat, where the work is not to be finished white, an approach must be made to the required colour.

The third coat is usually the last, and is made with a base of white lead, mixed with the requisite colour. and diluted with one-third of linseed oil to two-thirds of turpentine.

Painting on stucco, and all other work in which the surface is required to be without gloss, has an additional coat mixed with turpentine only, which, from its drying of one uniform *flat* tint, is called a flatting coat.

If the knots show through the second coat, they must be carefully covered with silver leaf.

Work finished as above described would be technically specified as knotted, primed, painted 3 oils, and flatted

Flatting is almost indispensable in all delicate interior work, but it is not suited to outside work, as it will not bear exposure to the weather

263. Painting on stucco is primed with boiled linseed oil, and should then receive at least three coats of white lead and oil, and be finished with a flat tint. The great secret of success in painting stucco is, that the surface should be perfectly dry; and, as this can hardly be the case in less than two years after the erection of a building, it will always be advisable to finish new work in distemper, which can be washed off whenever the walls are sufficiently dry to receive the permanent decorations.

264. *Graining* is the imitation of the grain of various kinds of woods, by means of *graining tools*, and, when well executed, and properly varnished, has a handsome appearance, and lasts many years. The term graining is also applied to the imitation of marbles.

265. Clear coling (from *claire colle*, i.e. transparent size, Fr.), is a substitution of size for oil, in the preparation of the priming coat. It is much resorted to by painters, on account of the ease with which a good face can be put on the work with fewer coats than when oil is used; but it will not stand damp, which causes it to scale off, and it should never be used except in repainting old work, which is greasy or smoky, and cannot be made to look well by any other means.

266. *Distempering* is a kind of painting in which whiting is used as the basis of the colours, the liquid medium being size; it is much used for ceilings and walls, and always will require two, and sometimes three coats, to give it a uniform appearance.

267. Painters' work is valued per superficial yard, according to the number of coats, and the description of work, as common colours, fancy colours, party colours, &c.

Where work is cut in on both edges, it is taken by the lineal foot In measuring railings, the two sides are measured as flat work. Sash frames are valued per piece, and sashes at per dozen squares

268. The manufacture of scagliola, or imitation marble, is a branch of the decorator's business, which is carried to very great perfection.

Scagliola is made of plaster of Paris and different earthy colours, which are mixed in a trough in a moist state, and blended together until the required effect is produced, when the composition is taken from the trough, laid on the plaster ground, and well worked into it with a wooden beater, and a small gauging trowel. When quite hard, it is smoothed, scraped, and polished, until it assumes the appearance of marble.

Scagliola is valued at per superficial foot, according to the description of marble imitated, and the execution of the work.

269. Gilding is executed with leaf gold, which is furnished by the gold-beater in books of 25 leaves, each leaf measuring $3\frac{1}{4}$ in. by 3 in. The parts to be gilded are first prepared with a coat of gold size, which is made of Oxford ochre and fat oil.

270. The operations of the paper-hanger are too simple to require description.

A piece of paper is 12 yards long, and is 20 in. wide, when hung, and covers 60 ft. superficial; hence the number of superficial feet that have to be covered, divided by 60, will give the number of pieces required.

Paper-hangers' work is valued at per piece, according to the value of the paper

The trades of the plumber, glazier, painter, paper-hanger, and decorator are often carried on by the same person

---

SECTION V.

WORKING DRAWINGS, SPECIFICATIONS, ESTI-MATES, AND CONTRACTS.

271. The erection of buildings of any considerable magnitude is usually carried on under the superintendence of a professional architect, whose duties consist in the preparation of the various working drawings and specifications that may be required for the guidance of the builder; in the strict supervision of the work during its progress, to insure that his instructions are carried out in a satisfactory manner; and in the examination and revision of all the accounts connected with the works.

This brief enumeration of the duties of an architect will suffice to show how many qualifications are required in one who aims at being thoroughly competent in his profession. He must unite the taste of the artist with the science and practical knowledge of the builder, and must be at the same time conversant with mercantile affairs, and counting-house routine, in order that he may avoid involving his employer in the trouble and expense attendant on disputed accounts, which generally are the result of the want of a clear and explicit understanding, on the part of the builder, of the obligations and responsibilities of engagements based upon the incomplete drawings or vaguely-worded specifications of an incompetent architect.

272. The profession of the architect and the trade of the builder are sometimes carried on by the same person: but this union of the directive and executive functions is not to be recommended; in the first place, because the duties of the workshop and the builder's yard leave little

time for the study of the higher branches of architectural knowledge; and, in the second place, because the absence of professional control will always be a strong temptation to a contractor to prefer his own interests to those of his employer, however competent he may be to design the buildings with the execution of which he may be charged.

During the present century, the impulse given to our arts and manufactures, and the improvements effected in the internal communications of the country, have given rise to the execution of many extensive works requiring for their construction a large amount of mechanical and scientific knowledge; in consequence of which a new and most important profession has sprung up during the last thirty years, occupying a middle position between those of architecture and mechanical engineering, viz., that of the civil engineer. The practice of the architect and of the civil engineer so closely approximate in many respects, that it is difficult strictly to draw the line of demarcation between them; but it may be said in general terms that, whilst the one is chiefly engaged in works of civil and decorative architecture, such as the erection of churches, public buildings, and dwelling-houses, the talent of the other is principally called forth in the art of construction on a large scale, as applied to retaining walls, bridges, tunnels, lighthouses, &c., and works connected with the improvements of the navigation and internal communications of the country.

273. The business of the surveyor is often carried on as a distinct branch of architectural practice; and, as the title of surveyor is often appropriated by those who have no real claim to it, a few words on a surveyor's duties may not be here out of place.

Surveyors may be divided into three classes; land surveyors, engineering surveyors, and building surveyors.

The business of an engineering surveyor, as distinguished from that of a land surveyor, chiefly consists in the preparation of accurate plans, sections, and other data relative to the intended sites of large works, which may be required by the architect or engineer preparatory to making out his working drawings, and in conducting levelling operations for drainage works, canals, railways, &c.

The building surveyor prepares, from the drawings and specifications of the architect or the engineer, bills of quantities of intended works, for the use of the builder, on which to frame his estimates; and, in the case of contracts, these bills of quantities form the basis of the engagements entered into by the builder and his employer, the surveyor being pecuniarily answerable for any omissions. The surveyor is also employed in the measurement of works already executed or in progress; in the latter case, for the purpose of ascertaining the advances to be made at stated intervals, and is engaged generally in all business connected with builders' accounts.*

274. The following is the general routine of proceedings in the case of large works. It will readily be understood that in small works subdivision of labour is not carried to such an extent, the architect superintending the works himself, without the aid of a clerk of the works, and the builders taking out their own quantities.

I. The general design having been approved of, and the site fixed upon, an exact plan is made of the ground, the nature of the foundation examined, and all the levels taken that may be required for the preparation of the working drawings.

II. The architect makes out the working drawings, and draws up the specification of the work.

III. A meeting is held of builders proposing to tender for the execution of the proposed works, called either by public advertisement or private invitation, at which a surveyor is appointed in their behalf to take out the quantities. Sometimes two surveyors are appointed, one on the part of the builders, and one on the part of the

---

* See "Student's Guide for Measuring and Estimating Artificer's Work," 2nd edition, 8vo, 1853.

architect, who take out the quantities together, and check each other as they proceed.

IV. The surveyor having furnished each party proposing to tender, with a copy of the bills of quantities, the builders prepare their estimates, and meet a second time to give in their tenders, after which the successful competitor and the employer sign a contract, drawn up by a solicitor, binding the one to the proper execution of the works, and the other to the payment of the amount of their cost at such times and in such sums as may be set forth in the specification.

V. The work is then set out,* and carried on under the

---

* *On Setting-out Work.*—The determination of the exact position of an intended building being sometimes difficult to accomplish, a few remarks on the subject may be acceptable.

The setting out of the leading lines is simple enough on level ground, where nothing occurs to interrupt the view, or to prevent the direct measurement of the required distances; but to perform this operation at the bottom of a foundation pit, blocked up with balks and shores, and ankle-deep in slush, requires a degree of practice and patience not always to be met with. Let us take a simple case, such as the putting in the abutment

Fig. 108.

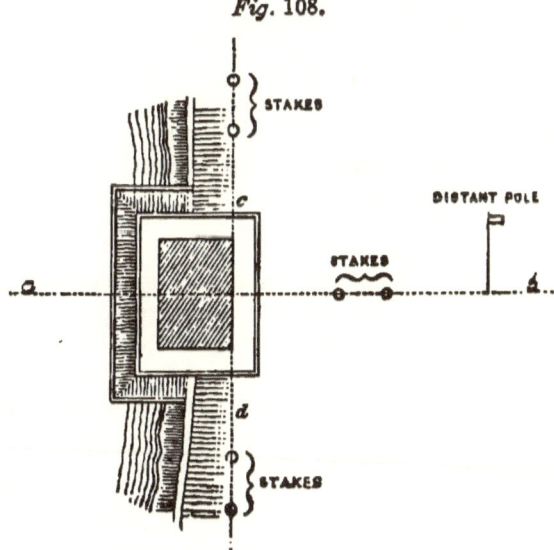

constant direction of a foreman on the part of the builder, and on the part of the architect under the superintendence of an inspector or clerk of the works, whose duty is to be constantly on the spot to check the quality and quantity of material used, to see to the proper execution of the work, and to keep a record of every deviation from the drawings that may be rendered necessary by the wishes of the employer, or by local circumstances over which the architect has no control

The work is measured up at regular intervals, and

---

of a bridge or viaduct, any error in the position of which would render the work useless (see fig. 108). The leading lines having been laid down on the drawings, the first thing to be done, before breaking ground, is to set out the centre line very carefully with a theodolite and ranging rods for a considerable distance on each side of the work, and to fix its position by erecting poles, planed true and placed perfectly upright, in some part of the line where there is no chance of their being disturbed.

Next, the exact position of the abutment on the centre line would be decided upon, and fixed by setting out another line at right angles to the first, as $c\,d$, which would also be extended beyond the works, and its position fixed by driving in stakes, the exact position of the line on the head

of the stake being marked by a saw-cut $\oplus$.

These guiding lines having now been permanently secured, the plan of the abutment may be set out on the ground, the dams driven, and the earth got out to the required depth. By the time the excavation is ready for commencing the work, it generally presents a forest of stays, struts, and shores, that would defy any attempt at setting out the work on its own level; it must, therefore, be set out at the level of the top or the dam, and the points transferred or *dropped* as follows:—

First, the position of the centre line is ascertained by reference to the poles, and, nails being driven into the timbers at the sides of the dam, a line line is strained across; the position of the line $c\,d$ is found, and a second line strained across in the same way. In a similar manner other lines are strained from side to side at the required distances, the lengths being measured from the line $c\,d$, and the widths from $a\,b$, until the outline of the foundation course is found; the angle points are then transferred to the bottom of the excavation by means of plumb-lines, and the work is commenced, its accuracy being easily tested by measurements from the lines $a\,b$ and $c\,d$, until it is so far advanced as to render this unnecessary.

payments made on account to the builder, upon the architect's certificate of the amount of work done.

VI. The work being completed, the extras and omissions are set against each other, and the difference added to or deducted from the amount of the contract, and the whole business is concluded by the architect giving a final certificate for the payment of the balance due to the builder

275. *Plan of Site.*—In preparing the plan of the site of any proposed works, the operations of the surveyor will generally have to be extended beyond the spot of ground on which the building is to stand. The frontages of the adjacent buildings, and the position of all existing or contemplated sewers, drains, and watercourses, should be correctly ascertained and laid down. Sketches drawn to scale of the architectural features of the adjacent buildings, if in town, and accurate outline sketches of the *incidents* of the locality of the intended operations, if in the country, should accompany the plan, that the architect may try the effect of his design before its actual execution renders it impossible to remedy its faults

By the careful study of all these data the architect may hope to succeed in making his works harmonise with the objects that surround them; without them, failure on this head is almost a certainty.

276. *Levels.*—Where the irregularities of the ground are considerable, it is necessary to ascertain the variations of the surface before the depth of the foundations and the position of the floors can be decided upon.

It also frequently happens that the levels of the floors and other leading lines, in a new building, are regulated by the capabilities of sewerage or drainage, or by the heights of other buildings with which the new work will ultimately be connected, as in the case of new streets. It therefore becomes of importance to have simple and accurate means of ascertaining and recording the relative heights of different points. For this purpose both the spirit level and the mason's level are used

277. Where the ground to be levelled over is limited in extent, and the variations of level do not exceed 12 feet the heights of any points may be found with the mason's level in the following manner. (*See* fig. 109.)

Fig. 109.

In a convenient place, near the highest part of the ground, drive three stout stakes at equal distances from each other, and nail to them three pieces of stout plank, placed as shown in the cut, their upper edges being adjusted to the same horizontal plane by means of the mason's level. The level being then placed on this frame, an assistant proceeds to the first point of which the height is required, holding up a rod with a sliding vane, which he raises or lowers in obedience to the directions of the surveyor, until it coincides with a pair of sights fixed at the bottom of the level; the height of the vane will then be the difference of level between the top of the levelling frame, and the place where the staff was held up

278. The above and similar methods will suffice for taking levels in a rough way for the ordinary purposes of the builder; but where great accuracy is requisite, or where the levels have to be extended over a considerable distance, as is often the case in drainage works, the use of a more perfect contrivance is necessary, and the spirit level is the instrument principally used for this purpose.

The spirit level consists of a telescope mounted on a portable stand, and furnished with screw adjustments, by means of which it can be made to revolve in a horizontal plane, any deviation from which is indicated by the motion of an air-bubble in a glass tube fixed parallel to the telescope.

The eye-piece of the telescope is furnished with cross-wires, as they are technically termed, made of spiders' thread, of which the use will be presently explained.

279. The levelling staff, now in common use, is divided into feet, tenths, and hundredths, in a conspicuous manner, so that, with the help of the glass, every division can be distinctly seen at the distance of one hundred yards or more. The mode of conducting the operation of levelling is as follows :—

The surveyor having set up and adjusted his instrument, the staffholder proceeds to the point at which the levels are to commence, and holds up his staff perfectly upright and turned towards the surveyor, who notes the division of the staff which coincides with the horizontal wire in the telescope, and enters the same in his level-book; the staff-holder then proceeds to the next point, and the reading of the staff is noted as before; and this is repeated until the distance or the difference of level makes it necessary for the surveyor to take up a fresh position. While this is being done, the staffholder remains stationary, until, the level being adjusted again, he carefully turns the face of the staff so as to be visible from the instrument in its new position, and a second reading of the staff is noted, after which he proceeds forward as before for a fresh set of observations.

280. In every set of observations the first is called a Backsight and the last a Foresight. The remaining observations are called intermediates, and are entered accordingly. It will be seen that an error in an intermediate reading is confined to the point where it occurs; but a mistake in a back or foresight is carried throughout the whole work, and therefore every care should be taken to insure accuracy in observing these sights.

281. The surveyor should commence and close his work by setting the staff on some well-defined mark, which can readily be referred to at any subsequent period, such as a door-step, plinth of a column, &c. These marks are called

bench marks, written B M, and are essential for either checking the work or carrying it on at a subsequent period.

282. The reduction of the levels to a tabular form for use is a simple arithmetical operation, which will be readily understood by examination of the annexed example of a level-book, and of the accompanying section,* fig. 110

Fig. 110.

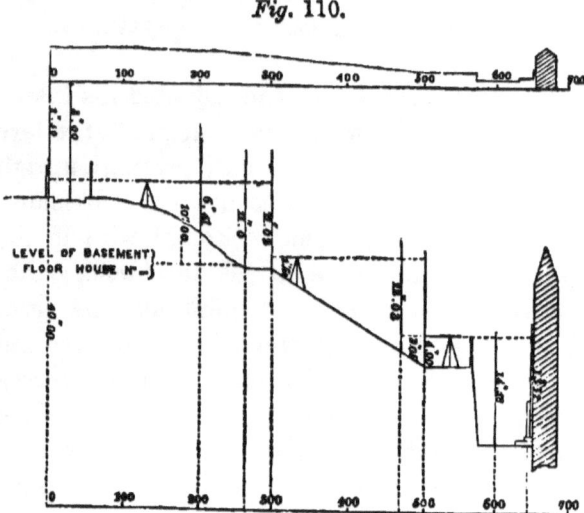

The difference between the successive readings in any set of observations is the difference of level between the points where the staff was successively held up, and by simple addition or subtraction, according as the ground rises or falls, we might obtain the total rise or fall of the ground above or below the starting point; but as this would require two columns, one for the total rise, and one for the total fall, it is simpler to assume the starting point to be some given height above an imaginary horizontal *datum line,*

---

* In plotting sections of ground, it is usual to make the vertical scale much greater than the horizontal, which enables small variations of level to be easily measured on the drawing without its being extended to an inconvenient length. This is shown in the lower half of fig. 110. The upper part of the figure shows the section plotted to the same horizontal and vertical scale.

drawn below the lowest point of the ground, to which level all the heights are referred in the column headed total height above datum line.

283. The accuracy of the arithmetical computations is proved by adding up the foresights and backsights, and, deducting the sum of the former from that of the latter (the height of the first B M having been previously entered at the top of the page as a backsight), the remainder will be the height of the last B M, and should agree with the last figures in the column of total heights.

284. In levelling the site of a proposed building, if no suitable object presents itself for a permanent B M for future reference, a large stake, hooped with iron, should be driven into the ground in some convenient place where it will not be disturbed. The height of this stake being then carefully noted and marked upon the elevations and sections of the building, it will serve as a constant check on the depths of the excavations and the heights of the different parts of the work, until the walls reach the level of the principal floor, when it will no longer be required.

285. We must not leave the subject of levels without mentioning a very useful instrument, called the water-level, which consists of a long flexible pipe, filled with water, and terminating at each end in an open glass tube. When it is required to find the relative heights of any two points, as, for instance, the relative levels of the floors of two adjoining houses, the two ends of the tube are taken to the respective points, the tube being passed down the staircases, over the roofs, or along any other accessible route, no matter how circuitous, and the required levels are found by measuring up from the floors to the surface of the water, which will of course stand at the same level at each end of the tube

WORKING DRAWINGS.

286 The architect being furnished with the plan and levels of the site of his operations, and having caused a

Readings of the Staff.			Difference of readings.		Reduced levels.	Distance in feet.	Remarks.
Back sight.	Inter-mediate.	Fore sight.	Rise.	Fall.	Total height above datum line.		Levels of Building Ground at —— (Date).
2.50	...	...	...		40.00	—	B M on doorstep of garden, No. — Park Road.
	3.00	...	...	0.50	39.50	30	On centre of Park Road.
	10.00	...	...	7.00	32.50	..	Level of basement floor at No. — * * * Terrace.
	11.00	...	...	1.00	31.50	260	Terrace walk.
1.60		11.02	...	0.02	31.48	300	
2.05		12.03	...	10.43	21.05	475	
	4.00	...	...	1.95	19.10	500	
	14.18	..	...	10.18	8.92	600	Centre of Lower Road.
		13.02	0.86	...	9.78	..	B M top of doorstep, No. — Lower Road.

40.00 Height of 1st B M above datum.

45.15
26.37
36.37

9.78 Reduced level of last B M.

careful examination to be made of the probable nature of the foundation by digging pits or taking borings, proceeds to make out his working drawings.

It is not sufficient for the execution of working drawings that the draughtsman should be acquainted with the ordinary principles of geometrical projection. He must also be thoroughly conversant with perspective, and with the principles of chiascuro, or light and shade, or he will work at random, as the geometrical projections which are required for the use of the workmen give a very false idea of the effect the work will have in execution.

287. Working drawings may be divided under three heads, viz. :—Block plans, General drawings, and Detail drawings :—

I. *Block Plans.*—These show the outline only of the intended building, and its position with regard to surrounding objects. They are drawn to a small scale, embracing the whole area of the site, and on them are marked the existing boundary walls, sewers, gas and water mains, and all the new walls, drains, and water-pipes, and their proposed connection with the existing ones, so that the builder may see at a glance the extent of his operations.

A well-digested block plan, with its accompanying levels, showing the heights of the principal points, the fall of the drains, &c., is one of the first requisites in a complete set of working drawings.

II. *General Drawings.*—These show the whole extent of the building, and the arrangement and connection of the different parts more or less in detail, according to its size and extent. These drawings consist of *Plans* of the foundations, and of the different stories of the building, and of the roofs; *Elevations* of the different fronts; and *Sections* showing the heights of the stories, and such constructive details as the scale will admit of. These drawings are carefully figured, the dimensions of each part being calculated, and its position fixed by reference to some well-defined line in the plans or elevations, the position of which admits of easy verification in all stages of the work This is best done by ruling faint lines on the drawings, through the principal divisions of the design, as shown in fig. 111, where the plan and elevation are divided into com·

Fig. 111.

partmonts by lines passing through the contres of the columns from which all the dimensions are dated each way These centre lines are, in the execution of the work, kept constantly marked on the walls as they are carried up, so that they are at all times available for reference.

By this means, the centre lines having been once carefully marked on the building, any slight error or variation from the drawings is confined to the spot where it occurs, instead of being carried forward, as is sometimes the case, to appear only when correction is as desirable as it is impossible.

The use of these centre lines also saves much of the labour of the draughtsman, as they form a skeleton, of which only so much need be filled up as may be required to show the design of the work.

III. *Detailed Drawings.*—These are on a large scale, showing those details of construction which could not be explained in the general drawings, such as the framing of floors, partitions, and roofs, for the use of the carpenter; the patterns of cast-iron girders and story posts for the iron-founder; decorative details of columns, entablatures, and cornices, for the carver; the requisite details being made out separately, as far as possible, for each trade; which arrangement saves much time that would otherwise be wasted in referring from one drawing to another, and, which is still more important, insures greater accuracy, from the workman understanding better the nature of his work.

In making the detailed drawings, every particular should be enumerated that may be required for a perfect under-standing of the nature and extent of the work. Thus, in preparing the drawings for the iron-founder, every separate pattern should be drawn out, and the number stated that will be required of each.

This principle should be attended to throughout the whole of the detailed drawings, as, in the absence of such data, it is very difficult to prepare correct estimates for the execution of the work without devoting more time to the study of the drawings than can generally be obtained for that purpose.

SPECIFICATION

288. The drawings being completed, the architect next draws up the specification of the intended works. This is divided under two principal heads—1st, the conditions of the contract; and 2nd, the description of the work.

The title briefly states the nature and extent of the works to be performed, and enumerates the drawings which are to accompany and to form part of the written specification.

289. *Conditions of Contract.*—Besides the special clauses and provisions which are required by the particular circum-stances of each case, the following clauses are inserted in all specifications :—

1. The works are to be executed to the full intent and

meaning of the drawings and specification, and to the satis
faction of the architect.

2. The contractor to take the entire charge of the works
during their progress, and to be responsible for all losses
and accidents until their completion.

3. The architect is to have power to reject all improper
materials or defective workmanship, and to have full control over the execution of the works, and free access at all
times to the workshops of the contractor where any work is
being prepared.

4 Alterations in the design are not to vitiate the contract,
but all extra or omitted works are to be measured and valued
according to a schedule of prices previously agreed upon.

5. The amount of the contract to be paid by instalments
as the works proceed, at the rate of — per cent. on the
amount of work done, and the balance within ——— from
the date of the architect's final certificate.

Lastly. The works are to be completed within a stated
time, under penalties which are enumerated.

290. *The description of the works* details minutely the quality of the materials, and describes the manner in which
every portion of the work is to be executed, the fulness of
the description depending on the amount of detailed information conveyed by the working drawings, care being taken
that the drawings and specification should, together, contain every particular that is necessary to be known in order
to make a fair estimate of the value of the work

291. The chief merit of a specification consists in the
use of clear and explicit language, and in the systematic
arrangement of its contents, so that the description of each
portion of the work shall be found in its proper place ; to
facilitate reference, every clause should be numbered and
have a marginal reference attached, and a copious index
should accompany the whole.

BILLS OF QUANTITIES

292 The surveyor being furnished by the architect with

the drawings and specification, proceeds to take out the quantities for the use of the parties who propose to tender for the execution of the work. This is done in the same way that work is measured when executed, except that the measurements are made on the drawings with a scale instead of on the real building with measuring rods.

293. In taking out quantities there are three distinct operations: 1st, taking the dimensions of the several parts of the work and entering them in the dimension book; 2ndly, working out the quantities from the dimensions, and posting them into the columns of the abstracts, which is called *abstracting;* 3rdly, casting up the columns of the abstracts and bringing the quantities into bill.

294. The dimension book is ruled and the dimensions entered as in the following examples:—

No.	Dimension.	Quantity.	Description.
16	ft. in. 14 0 0 10 0 2½	ft. in. 38 10	Memel fir framed joists to front room ground floor.

In this example, the work measured consists of sixteen joists, each 14 ft. long and 10 in. deep and 2½ in thick, and the total quantity of timber they contain amounts to 38 ft 10 in. cube.

Dimension.	No. of bricks in thickness.	Quantity.	Description.
ft. in. 20 6 11 6	2½	ft. in. 235 9	Stock brickwork in mortar to front wall, from footings to 1st set-off.

This example needs no explanation.

295. In preparing the abstract for each trade, the surveyor looks over his dimensions to see what articles he will

have, and rules his paper into columns accordingly, writing the proper heads over each.

The principal point to be attended to in abstracting quantities is, to preserve a regular rotation in arranging the different descriptions of work, so that every article may at once be found on referring to its proper place in the abstract.

No fixed rules can be given on this head, as the form of abstract is different for every trade, and must be varied according to circumstances; but, as a general principle, articles of least value should be placed first. Solid measure should take precedence of superficial, and superficial of lineal, and miscellaneous articles should come last of all; or in technical terms, the rotations should be, 1st, cubes, 2nd, supers.; 3rd, runs; and, lastly, miscellaneous.

296. In bringing the quantities into bill, the same rotation is to be observed as in abstracting them, care being taken that every article is inserted in its proper place, so that it may readily be found in the bill.

The limits of this volume prevent our going into much detail on the subject of builders' accounts, and we must therefore confine ourselves to laying before the reader a skeleton estimate, which will give him a tolerable idea of the manner in which the several kinds of artificers' work are abstracted and brought into bill.

297. Estimate for the Erection of ——— at ——— for ———, according to Specification and Drawings numbered 1 to —, prepared by ——— Architect.

(Date)

FOUNDATIONS.

y'ds.	ft.			£	s.	d.
—	— cube	Excavation to foundations (including cofferdams, pumping, &c. . . .	at —	—	—	—
—	— „	Concrete	„ —	—	—	—
ft.	**in.**					
—	— „	Timber in piles driven — ft. through (describe the material), including ringing, shoeing, and driving, but not ironwork	„ —	—	—	—
—	— „	Do. in 6-in. planking, spiked to pile-heads	„ —	—	—	—

Carried forward . . . £— — —

FOUNDATIONS *continued.*

cwt.	qrs	lbs.	Brougnt forward .		£	s.	d.
—	—	—	Wrought iron in shoes to piles . .	at —	—	—	—
			Total of foundations to be carried to summary . . .		—	—	—

BRICKLAYER.

rods	ft.				£	s.	d.
—	—	supl.	Reduced brickwork in mortar . .	„ —	—	—	—
—	—	„	Do. do. in cement . .	„ —	—	—	—
sqrs.	ft.						
—	—	„	Tiling (describing the kind, whether plain or pantiling, if single or double laths, &c., &c.)	„ —	—	—	—
yds.	ft.						
—	—	„	Bricknogging to partitions . .	„ —	—	—	—
—	—	„	Paving (of various descriptions) .	„ —	—	—	—
—	—	„	And all other articles valued per yard superficial.				
ft.	in.						
—	—	„	Gauge arches	„ —	—	—	—
—	—	„	Facings (with superior description of bricks, specifying the quality) .	„ —	—	—	—
—	—	„	Cutting to arches or splays . .	„ —	—	—	—
			And all other work valued by the foot superficial.				
—	—	run	Barrel or other drains (specifying size, &c.)	„ —	—	—	—
—	—	„	Tile creasing	„ —	—	—	—
			And all other articles valued by running measure.				
		Nos.	Chimney pots, each; bedding and pointing sash and door frames, each; and all miscellaneous articles .		—	—	—
			Total of bricklayers' work to be carried to summary . . .		—	—	—

MASON.

yds.	ft.				£	s.	d.
—	—	cube	Rubble walling	„ —	—	—	—
—	—	„	Hammer-dressed walling in random courses	„ —	—	—	—
ft.	in.						
—	—	„	Stone (describing the kinds) . .	„ —	—	—	—
—	—	supl.	Labour on above (as plain work, sunk, moulded or circular work) . .	„ —	—	—	—
—	—	„	Hearths, pavings, landings, &c., beginning with the thinnest . .	„ —	—	—	—
—	—	„	Marble slabs, beginning with the thinnest and inferior qualities .	„ —	—	—	—
—	—	run	Window sills, curbs, steps, copings, &c.		—	—	—
—	—	„	Joggle joints, chases, &c. . .	„ —	—	—	—

Carried forward . . . £ — — —

MASON *continued.*

			£	s.	d.
	Brought forward . . .				
Nos.	Mortices and rail holes, &c.—dowels, cramps, and other articles numbered		—	—	—
	Total of mason's work to be carried to summary	£	—	—	—

CARPENTER AND JOINER.

sqrs.	ft.			£	s.	d.
—	—	supl.	Labour and nails to roofs, floors, or quarter partitions . . .	at—	—	—
—	—	„	Battenings and boardings according to description	„ —	—	—
--	—	„	Floors, according to description, beginning with the inferior and ending with the best descriptions . . And so on for all work valued by the square.	„ —	—	—

ft.	in.					
—	—	cube	Memel fir, according to description, as fir bond, fir framed, wrought and framed, wrought, framed, and rebated, &c.	„ —	—	—
—	—	„	Do. proper door and window cases . Then oak and superior descriptions of timber, in the same way. Then the superficial work, as—	„ —	—	—
—	—	supl.	½-in. deal rough linings, and so on with the different thicknesses of deals according to the labour on them; arranging them according to their thickness and the amount of labour on them, beginning with the thinnest . . . . . Then oak plank or mahogany in the same way. Then take the framed work, as—	„ —	—	—
—	—	„	1¼-in. deal square-framed inclosure to closets, and so on with the rest of the framed work, as doors, shutters, sashes, frames, &c., according to description . . . . . Then the work valued by running measure, as—	„ —	—	—
—	--	run	2¼-in. Spanish mahogany moulded, grooved, and beaded handrail . Then the numbers, as—	„ —	—	—
Nos.			Mitred and turned caps, fixing iron balusters, &c.	„ —	—	...
			Lastly — The Ironmongery, every article of which should be carefully described	„ —	—	—
			Total of carpenter and joiner's work to be carried to summary .	£	—	—	—

			£	s.	d.
sqrs. ft. — — supl.	**SLATER.** Countess, or any other kind of slating, according to description . . at —		—	—	—
ft. in. — — „	Then slate slab, as— Inch shelves, rubbed one side, beginning with the slabs of least thickness, and arranging them according to the labour bestowed on them .	— „	—	—	—
— — run	Then the work valued by running measure, as— Patent saddle-cut slate ridge . .	— „	—	—	—
Nos.	Lastly the numbers, as— Holes, cut, &c.	— „	—	—	—
	Total of slater's work to be carried to summary	£	—	—	—

PLASTERER.

			£	s.	d.
yds. ft. — — supl.	First the superficial quantity of plastering, as— Render float and set to walls, beginning with the commonest, and proceeding through the different descriptions of two and three coat work up to the stuccos and superior work	— „	—	—	—
	Then the whitewashing, distempering, &c.	— „	—	—	—
ft. in. — — run	Next the run of cornices, architraves, reveals, &c., as— Plain cornice to drawing-room, 14-in. girt	— „	—	—	—
Nos.	And lastly the numbers, as— 4 mitres, 1 centre flower, 30 in. diameter, &c., &c.		—	—	—
	Total of plasterer's work to be carried to summary . . .	£	—	—	—

SMITH AND IRON-FOUNDER.

			£	s.	d.
tons.cwt.qrs.lbs — — — —	Begin with the cast-iron, as— Cast iron in No. 4 girders, including patterns, painting, and fixing . N.B.—State the No. of patterns.	— „	—	—	—
	Then the smaller castings, as— Railings, balconies, columns, &c. .	— „	—	—	—
— — — —	Then the wrought iron, as— Wrought iron in chimney bars, straps, screw bolts, railings, &c. . .	— „	—	—	—
yds. ft. — — run	Then the articles sold by running measure, as— Cast-iron gutters, water-pipes, &c. .	— „	—	—	—
	Carried forward . . .	£	—	—	—

SMITH AND IRON-FOUNDER *continued.*

		£	s.	d.
Brought forward . . .		—	—	—
Lastly the numbers, as—				
Nos. Stoves, coal-plates, stable-fittings, &c. .		—	—	—
Total smith and iron-founder's work				
to be carried to summary . . £		—	—	—

BELL-HANGER.

		£	s.	d.
Number the bells, and describe the mode of hanging, as—				
Nos. — bells hung with copper wires in concealed tin tubes, with bells, cranks, and wires complete . .	„ —	—	—	—
And then enumerate the ornamental furniture to the different pulls .	„ —	—	—	—
Total of bell-hanger's work to be carried to summary . . . £		—	—	—

PLUMBER.

cwt. qrs. lbs.

		£	s.	d.
— — — Cast lead laid in gutters, hips, ridges, flats, cisterns, &c.; including all solder, wall hooks, nails, &c. .	„ —	—	—	—
— — — Milled do. do. .	„ —	—	—	—
Then socket, rain-water, and funnel pipes, and other work valued by				
ft. in. the lineal foot, as—				
— — run Inch drawn pipes	„ —	—	—	—
Lastly the numbers, as—				
Nos. Joints, plugs, and washers, air traps, brass grates, cocks, copper balls, pumps, water closets, apparatus, &c.				
Total of plumber's work to be carried to summary . . . £		—	—	—

PAINTER.

yds. ft.

		£	s.	d.
— — supl. Of painting, according to description, specifying the number of oils, and whether common or extra colours, beginning with the work in fewest coats and finishing with the most expensive descriptions . . .	„ —	—	—	—
ft. in. Then the running work, as—				
— — run Skirtings, plinths, window sills, &c.	„ —	—	—	—
Lastly the numbers, as—				
Nos. Frames, squares, chimney pieces, &c. .	„ —	—	—	—
Total of painter's work to be carried to summary	£	—	—	—

ft. in.	GLAZIER.		£	s.	d.
— — supl.	Glazing, according to description, specifying size of squares and quality of glass . . . . at	—	—	—	—
	Then the stained and other ornamental glass; and, lastly, the plate glass.				
	Total of glazier's work to be carried to summary	£	—	—	—

	PAPER-HANGER AND DECORATOR.		£	s.	d.
yds. ft.					
— — supl.	Distempering, according to description	„ —	—	—	—
ft. in.					
— — „	Scagliola slabs do. . .	„ —	—	—	—
yds. ft.					
— — run	Gold mouldings	„ —	—	—	—
Nos.	Pieces of paper hung, according to description, including preparing walls	„ —	—	—	—
	—Hanging, lining, paper, and pumicing do.	„ —	—	—	—
„	Dozen of borders . . .	„ —	—	—	—
	Total of paper-hanger and decorator's works to be carried to summary	£	—	—	—

SUNDRIES.		£	s.	d.
Temporary fencings—watching and lighting works		—	—	—
Office for clerk of works		—	—	—
District surveyor's fee		—	—	—
Fire insurance		—	—	—
Surveyor's charge for bills of quantities . .		—	—	—
Total sundries to be carried to summary . .	£	—	—	—

SUMMARY OF BILLS.		£	s.	d.
Foundations		—	—	—
Bricklayer		—	—	—
Mason		—	—	—
Carpenter and joiner		—	—	—
Slater		—	—	—
Plasterer		—	—	—
Smith and iron-founder		—	—	—
Bell-hanger		—	—	—
Plumber, painter, and glazier . . .		—	—	—
Paper-hanger and decorator . . .		—	—	—
Sundries		—	—	—
Total amount of estimate.	£	—	—	—

298. The surveyor furnishes the builder, whose tender is accepted, with copies of the drawings from which the quantities have been taken off.

By reference to these, the builder can at all times satisfy himself that the detailed drawings, furnished for the execution of the work, contain nothing beyond what he has contracted for.

Copies of the drawings and specification are attached to the contract deed, and are signed by the builder and other parties respectively concerned.

299. It scarcely ever happens that a large undertaking can be carried into execution without considerable departure from the contract designs, especially in the matter of foundations and underground work; the exact nature and extent of which must often be uncertain until the works are commenced.

To provide for these contingencies without setting aside the contract, the builder's estimate is accompanied by a schedule of prices at which he undertakes to execute any additional work that may be required, or to value any work that may be omitted. This schedule should be very carefully drawn out, so that there shall be no dispute as to its meaning; thus, under the head of brickwork, it should be clearly understood whether centering is included in the price named, or whether it is to form an additional charge; with iron-founder's work, whether the price includes patterns, and so on with every description of work

300. For taking out quantities, surveyors are allowed a commission of $2\frac{1}{2}$ per cent. on the cost of the work, and they are responsible to the builder for any omissions which may have to be made good by the latter.

301. Architects are remunerated by a commission of 5 per cent. on the amount expended under their direction, besides travelling expenses, salary of the clerk of the works, and occasionally other charges, according to circumstances

APPENDIX OF NOTES AND ILLUSTRATIONS.

Note A, Page 13.—RETAINING WALLS.

The author, in the preface to the earlier editions, stated that authors prior to that date, from having neglected to take into consideration the friction of earth or gravel upon itself, had rendered the formulæ at which they had arrived, giving the conditions of stability of retaining walls, &c., of little or of no value in practice. He quotes Mr. Gwilt, who, in his "Encyclopædia of Architecture," Article 1,584, states that "he leaves out of question the rules of Dr. Hutton, as being absurd and incomprehensible;" but rightly adds that Hutton's formulæ, upon Hutton's data, are correct, and only require the correction for friction to make them agree with modern practice.

We need not, in a work so elementary as this, discuss that question here. The author, however, wrote at a period anterior to the publication of the formulæ and rules given by Professor Moseley, in his "Engineering and Architecture," and by Professor Rankine, in his volume of "Mechanics applied to Civil Engineering."

To both of these the student may be referred who desires thoroughly to understand the subject generally; the more advanced student will find a mass of important matter, theoretical and experimental, scattered through English and foreign engineering literature, and will derive great advantage by studying the examples with the conditions in which they were produced by some of the more celebrated engineers at home and abroad.

Note B, Page 30.—EASING DOWN CENTERING.

Since the date of the earlier editions of this work two methods of easing down centering—both, however, proposed at a considerably anterior period—have been brought into use with success and advantage. The first of these consists in substituting for the great chase wedges, and placed in the same position in which those are shown in Fig. 25, short but strong and powerful screw-jacks. By these the easing down is effected without the necessity for those tremendous

blows from a battering-ram necessary to start large chase wedges, and
the danger of the wedges becoming so set together that they cannot be
started at all, which has ere now happened, is avoided.   The second
method consists in supporting the centering at like points to the above
by shallow iron, or strong timber, boxes filled with *dry* sand, which is
permitted to run out by a lateral aperture, like sand from a common
hour-glass, when the lowering is to occur.   This has been found to
answer well on the Continent, and also in India, in respect of some of
the large viaduct arches upon the Great East Indian Railway.

Note C, Page 42.—FIRE-PROOF FLOORS.

The student will do well to make clear his notions as to what are the
conditions requisite in a floor, still more in an entire building, that it
shall be rightly entitled *Fire-proof.*   The vast mass of so-called floors
and buildings are mere deceptive shams.

In the loose language of every day converse, as well as in much that
in type ought to be more precise, anything is called a fire-proof floor
or building, provided well-known combustible materials, such as
timber, does not enter into the skeleton of the structure. The word is even
rather audaciously applied to structures such as the so-called fire-proof
flooring of Fox and Barrett's patent, in which a very considerable proportion of timber is employed, though shut up from the eye, and, we
may add, from the free air, which, as Mr. Dobson remarks, is essential
to its durability.   Neither these floors nor those so extensively
employed in the north of England and in Scotland for factory floors, as
shown in section in Fig. 50, nor a multitude of other constructions,
patented or otherwise, are really fire-proof at all.   The most that can
be said for them is that *they may delay* the spread of conflagration.

Upon the subject of fire-proof construction the student should consult various papers in the " Minutes of Proceedings of the Institution
of Civil Engineers," and in the leading technical journals, and the
works on the subject of the late Mr. Braidwood, of the London Fire
Brigade, and of Mr. Young, recently published.

Note D, Page 47.—ROOFS OF LARGE SPAN.

A narrow limit is set to the clear span of roofs when they are constructed of timber, as will be readily seen by the student who has
mastered properly the elements of physics as applied to structure.

The modulus of compression of all such timber as can be commanded
in sufficient quantity in the necessary long and straight lengths, and
at a cheap enough rate, is so low that the crushing up, both laterally
and in the end way of the grain, fixes their limit.   The use of iron
has, however, given an immense extension to the powers of the engineers in producing roofs of vast clear span.   Two hundred and fifty
feet, which, unless at prodigious cost, approaches the practical limit of
span in timber roofs with any permanent covering heavier than sheet

copper or zinc, is but the starting-point for future great roofs in iron, or still more in steel.

There can be no doubt that roofs of such material of 1,000 feet span may, with perfect ease and without any prohibitory cost, be safely produced.

And if full advantage be taken in the structural details of the tensional resistance of steel bars, probably double that span, if ever required, would not prove too heavy a tax upon the existing resources of the engineer. The iron roof of the St. Pancras Station, London, of the Midland Railway, now in progress, from the design of P. W. Barlow, Esq., C.E., is probably the largest in span yet attempted.

Note E, Page 59.—ZINC COATING, &c

The coating of wrought-iron and cast-iron with zinc, or with some alloys of zinc and lead, or zinc and tin, into which, when in fusion, the iron is dipped with certain preparations and precautions, has become extensively practised since M. Sorel, a Frenchman, first proposed the process, which has been the subject of many inferior British patents now expired. As a method of protection against the weather, *i.e.* against the conjoint effects of air and moisture, coating with *pure* zinc, if thoroughly well done, is tolerably effectual when applied to *wrought* iron. It is never so when applied to cast-iron. Nor is it, as a protection, of the slightest value upon either wrought or cast-iron, if these are exposed to the atmosphere of our coal-burning cities. The cast-iron plates of the roof of the Houses of Parliament present a lamentable proof of this fact, and wrought iron equally acted on may be seen in numberless places in London and other English towns.

In the pure atmosphere of the country—in agricultural districts and those not too near the sea, where the saline spray carried in by storms acts upon it—zincked or galvanised iron, as a covering, may be trusted, especially in dryer climates than ours, such as Canada or Australia.

Zinc itself, even the purest, such as that supplied by Mossilman, or the Vieulle Montagne Company, is not proof against the corrosive action of the smoke and sulphur-laden atmosphere of our coal-burning towns and manufacturing districts.

Upon the whole subject of the action of air and water upon the metals employed in the construction, the student will consult with advantage the four reports of Mr. Robert Mallet, prepared by the desire of the British Association for the Advancement of Science, and published in the volume of reports of that body.

Note F, Page 62.—HEATING AND VENTILATION.

The student who desires to make himself well acquainted both with the theory and best practice as respects the heating and ventilation of buildings, should study the recently published work by General Morin —" Etudes sur la Ventilation," 2 vols., 8vo., Paris. In no single work will be found so much and such sound and reliable information; but the subject is a very wide one, and no architectural student can be deemed instructed thoroughly in this important branch of his profession who has not consulted a very wide range of literature relating to it. A few of these treatises we may especially refer him to :—Tredgold on Heating and Ventilation, now getting a little antiquated; the Reports on the Heating and Ventilation of Pentonville and other Model Prisons; those on the same in relation to the Houses of Parliament, and those of the commission appointed to inquire into the means of heating and ventilating barracks and other apartments of relatively small size.

Note G, Page 68.—Preservation of Timber.

In addition to the several processes described in the text, the student should also be referred to the method of M. Boucherie, of at once seasoning and securing against decay timber in its green state, by causing the still living timber to take up by its capillary vessels a solution of sulphate of copper ; and a still more recent French invention, consisting in preventing the external attack of those fungous sporules which initiate decay, by superficially *charring* to an extremely small depth the whole of the surfaces of finished timber, *i.e.* the wood cut to the size, &c.

This is effected by rapidly passing over it a powerful flame produced by a coal-gas and air blow-pipe, or by the flame driven forth by air-blast from a coke fire in a small hollow furnace. Accounts of both these processes may be found in the transactions and journals of a technical class during the last few years.

Note H, Page 75.—Steel Manufacture.

Since the date of the original edition of this volume, several new and important processes for the production of steel have been invented and brought into practical use. For a full account of these the student should refer to Percy's "Metallurgy of Iron," or other recent systematic works on metallurgy.

The only two to which we need here briefly refer are the production of steel by a slight modification of the puddling process, by which ordinary wrought iron is produced. The steel thus obtained is called "puddled steel," and that procured by Mr. Bessemer's process, or by blowing common air through cast iron in a state of fusion. The cast iron must either contain some manganese, or iron containing that in alloy must be added in fusion during the process of blowing through to produce good steel. By either of the above processes excellent steel of various qualities, some equal to the best cast steel of cementation, may be produced.

Both processes have already greatly reduced the price and increased the supply of steel, and the Bessemer process seems almost certain to revolutionise, before long, the whole iron trade, as well as all the arts of construction, more or less, in proportion as they are dependent at present on wrought or cast iron.

Note I, Page 77.—CAST IRON, &c.

Although what the author has stated on the subject of the classes, properties, and manipulation of cast irons is well founded, his treatment of the subject, which is a large and complex one, is necessarily very incomplete. No student of the arts of construction, whether in the direction of engineering or of architecture, should deem himself educated until he has learnt enough of chemical metallurgy to read and make his own what has been written upon the physico-chemical history and proportions of iron in its three distinct states of cast iron, steel, and wrought iron—all highly complex bodies, and endowed with properties as remarkable as any in the range of entire nature. No single work will be found to give a clearer view of these properties and relations than Karsten's "Metallurgie di Fer," translated from the German, original by Kuhlmann (3 vols. 8vo). The student who once has mastered the scientific *principles* upon which the properties and changes in them of iron in its various states depend, will find himself in possession of a treasure that in practice will be as "a lantern to his feet" during his whole career in life, and, wanting which, all the random facts he may pick up out of popular text-books, journals, &c., will prove to him but a deceptive and disconnected jumble.

Note K, Page 80.—TENSILE RESISTANCE.

The table here given of the ultimate tensile resistances of these several materials must be accepted as only approximate and with caution.

No cast iron, but white mottled cast iron of great rigidity, will stand a tensile strain of $7\frac{3}{4}$ tons per square inch; nearly all the cast iron employed in architectural and engineering structures is torn asunder at a strain of four or five tons, and should never be trusted in tension at a strain of more than two tons per square inch for a dead pull, or one half of that for a sudden or impulsive strain.

Cast iron in different sized specimens of various qualities differs in tensile strength enormously—as much as 7 to 1.

It is not the popularly supposed *brittleness* of cast iron that makes it unsafe in tension, for within its range of resistance it is *less brittle* (*i.e.* more extensible or ductile) than wrought iron, but the small total range of its tensile resistance; whereas in wrought iron, this range being far larger, it constitutes for such moderate stress as that material should ever be exposed to in practice an actual margin of safety.

What the author meant by stating that *wrought iron* is torn asunder by ten tons and *English bar iron* by twenty-five tons, to the square inch of section, it is difficult to conjecture. Some British bar or plate irons of extreme softness and ductility, and therefore of extreme value for certain special purposes, are torn asunder by a strain of only about seven tons per square inch, while rigid and very little extensible bar iron occurs constantly which is not torn asunder under thirty tons per square inch. Steel, again, in bars, whether made by cementation, by puddling, or by the Bessemer process, can be produced with a tensile resistance of eighty tons or upwards per square inch, and the same material drawn into wire assumes a resistance of 120 tons per square inch.

The student must guardedly distinguish in every case, and above all in calculating dimensions to be given in practice to parts, whether to be exposed to tension, compression, or transverse strain, between *ultimate resistance* and *safe resistance;* what difference must be given under given conditions to the dimensions of the parts, or, what is the same thing, *what co-efficient*, as that of assumed ultimate resistance, he must employ, so as to ensure a sufficient margin or factor of safety, should be sought for by a careful study of a few of the most important of the mass of works treating of such subjects. Of these we may notice for his information the parliamentary "Report of the Commission on Railway Structures in Iron," Hodgkinson's edition of "Tredgold on Iron," Fairbairn's "Reports of Experiments to British Association," Edwin Clarke's "Britannia and Conway Bridges," and the able digests of the whole subject in his "Mechanics applied to Civil Engineering."

Mr. Kirkaldy's volume is valuable as an important contribution of facts, but the student should be warned to accept his generalisations with caution. Several of these are not in accordance with known facts of molecular physics, nor even with a correct interpretation of his own experiments, which are themselves creditable to his industry and zeal.

For the metallurgic properties of iron and steel exposed to strains, and especially as respects impulsive strains, Mr. Mallet's work on the "Materials employed in the Construction of Artillery" may be consulted.

Note L, Page 88.—STRAINS ON GIRDERS.

The margin of safety given in the text of two-thirds is, by consent of a large number of well-informed engineers, deemed too small— three-quarters factor of safety for *statical loads, i.e. dead weight*, is better and safer practice. Where the load is a rolling load, or variable, or liable by any chance to become a *dynamic load, i.e.* a *suddenly applied* or *impulsive load*, then double the above margin of safety should always be allowed, though it is not uncommon in English practice to allow for dynamic strain a section of resistance that shall reduce the strain upon the unit of section to only one-sixth that of the breaking strain.

INDEX.

THE END

PRINTED BY J. S. VIRTUE AND CO., LIMITED, CITY ROAD, LONDON.

WEALE'S RUDIMENTARY SCIENTIFIC SERIES.

*** The volumes of this Series are freely Illustrated with Woodcuts, or otherwise, where requisite. Throughout the following List it must be understood that the books are bound in limp cloth, unless otherwise stated; *but the volumes marked with a ‡ may also be had strongly bound in cloth boards for 6d. extra.*

N.B.—In ordering from this List it is recommended, as a means of facilitating business and obviating error, to quote the numbers affixed to the volumes, as well as the titles and prices.

CIVIL ENGINEERING, SURVEYING, ETC.

No.

31. *WELLS AND WELL-SINKING.* By JOHN GEO. SWINDELL, A.R.I.B.A., and G. R. BURNELL, C.E. Revised Edition. With a New Appendix on the Qualities of Water. Illustrated. 2s.

35. *THE BLASTING AND QUARRYING OF STONE,* for Building and other Purposes. By Gen. Sir J. BURGOYNE, Bart. 1s. 6d.

43. *TUBULAR, AND OTHER IRON GIRDER BRIDGES,* particularly describing the Britannia and Conway Tubular Bridges. By G. DRYSDALE DEMPSEY, C.E. Fourth Edition. 2s.

44. *FOUNDATIONS AND CONCRETE WORKS,* with Practical Remarks on Footings, Sand, Concrete, Béton, Pile-driving, Caissons, and Cofferdams, &c. By E. DOBSON. Fifth Edition. 1s. 6d.

60. *LAND AND ENGINEERING SURVEYING.* By T. BAKER, C.E. Fifteenth Edition, revised by Professor J. R. YOUNG. 2s.‡

80*. *EMBANKING LANDS FROM THE SEA.* With examples and Particulars of actual Embankments, &c. By J. WIGGINS, F.G.S. 2s.

81. *WATER WORKS,* for the Supply of Cities and Towns. With a Description of the Principal Geological Formations of England as influencing Supplies of Water, &c. By S. HUGHES, C.E. New Edition. 4s.‡

118. *CIVIL ENGINEERING IN NORTH AMERICA,* a Sketch of. By DAVID STEVENSON, F.R.S.E., &c. Plates and Diagrams. 3s.

167. *IRON BRIDGES, GIRDERS, ROOFS, AND OTHER WORKS.* By FRANCIS CAMPIN, C.E. 2s. 6d.‡

197. *ROADS AND STREETS.* By H. LAW, C.E., revised and enlarged by D. K. CLARK, C.E., including pavements of Stone, Wood, Asphalte, &c. 4s. 6d.‡

203. *SANITARY WORK IN THE SMALLER TOWNS AND IN VILLAGES.* By C. SLAGG, A.M.I.C.E. Revised Edition. 3s.‡

212. *GAS-WORKS, THEIR CONSTRUCTION AND ARRANGEMENT;* and the Manufacture and Distribution of Coal Gas. Originally written by SAMUEL HUGHES, C.E. Re-written and enlarged by WILLIAM RICHARDS, C.E. Seventh Edition, with important additions. 5s. 6d.‡

213. *PIONEER ENGINEERING.* A Treatise on the Engineering Operations connected with the Settlement of Waste Lands in New Countries. By EDWARD DOBSON, Assoc. Inst. C.E. 4s. 6d.‡

216. *MATERIALS AND CONSTRUCTION;* A Theoretical and Practical Treatise on the Strains, Designing, and Erection of Works of Construction. By FRANCIS CAMPIN, C.E. Second Edition, revised. 3s.‡

219. *CIVIL ENGINEERING.* By HENRY LAW, M.Inst. C.E. Including HYDRAULIC ENGINEERING by GEO. R. BURNELL, M.Inst. C.E. Seventh Edition, revised, with large additions by D. KINNEAR CLARK, M.Inst. C.E. 6s. 6d., Cloth boards, 7s. 6d.

268. *THE DRAINAGE OF LANDS, TOWNS, & BUILDINGS.* By G. D. DEMPSEY, C.E. Revised, with large Additions on Recent Practice in Drainage Engineering, by D. KINNEAR CLARK, M.I.C.E. Second Edition, Corrected. 4s. 6d.‡   [*Just published.*

☞ *The ‡ indicates that these vols. may be had strongly bound at 6d. extra.*

LONDON : CROSBY LOCKWOOD AND SON,

MECHANICAL ENGINEERING, ETC.

The ‡ indicates that these vols. may be had strongly bound at 6d. extra.

MINING, METALLURGY, ETC.

4. *MINERALOGY*, Rudiments of; a concise View of the General Properties of Minerals. By A. Ramsay, F.G.S., F.R.G.S., &c. Third Edition, revised and enlarged. Illustrated. 3s. 6d.‡

117. *SUBTERRANEOUS SURVEYING*, with and without the Magnetic Needle. By T. Fenwick and T. Baker, C.E. Illustrated. 2s. 6d.‡

135. *ELECTRO-METALLURGY*; Practically Treated. By Alexander Watt. Ninth Edition, enlarged and revised, with additional Illustrations, and including the most recent Processes. 3s. 6d.‡

172. *MINING TOOLS*, Manual of. For the Use of Mine Managers, Agents, Students, &c. By William Morgans. 2s. 6d.

172*. *MINING TOOLS, ATLAS* of Engravings to Illustrate the above, containing 235 Illustrations, drawn to Scale. 4to. 4s. 6d.

176. *METALLURGY OF IRON*. Containing History of Iron Manufacture, Methods of Assay, and Analyses of Iron Ores, Processes of Manufacture of Iron and Steel, &c. By H. Bauerman, F.G.S. Sixth Edition, revised and enlarged. 5s.‡   [*Just published.*]

180. *COAL AND COAL MINING*. By the late Sir Warington W. Smyth, M.A., F.R.S. Seventh Edition, revised. 3s. 6d.‡ [*Just published.*]

195. *THE MINERAL SURVEYOR AND VALUER'S COMPLETE GUIDE*. By W. Lintern, M.E. Third Edition, including Magnetic and Angular Surveying. With Four Plates. 3s. 6d.‡

214. *SLATE AND SLATE QUARRYING*, Scientific, Practical, and Commercial. By D. C. Davies, F.G.S., Mining Engineer, &c. 3s.‡

264. *A FIRST BOOK OF MINING AND QUARRYING*, with the Sciences connected therewith, for Primary Schools and Self-Instruction. By J. H. Collins, F.G.S. Second Edition, with additions. 1s. 6d.

ARCHITECTURE, BUILDING, ETC.

16. *ARCHITECTURE—ORDERS*—The Orders and their Æsthetic Principles. By W. H. Leeds. Illustrated. 1s. 6d.

17. *ARCHITECTURE—STYLES*—The History and Description of the Styles of Architecture of Various Countries, from the Earliest to the Present Period. By T. Talbot Bury, F.R.I.B.A., &c. Illustrated. 2s.
₊ Orders and Styles of Architecture, in One Vol., 3s. 6d.

18. *ARCHITECTURE—DESIGN*—The Principles of Design in Architecture, as deducible from Nature and exemplified in the Works of the Greek and Gothic Architects. By E. L. Garbett, Architect. Illustrated. 2s.6d.
₊ The three preceding Works, in One handsome Vol., half-bound, entitled "Modern Architecture," price 6s.

22. *THE ART OF BUILDING*, Rudiments of. General Principles of Construction, Materials used in Building, Strength and Use of Materials, Working Drawings, Specifications, and Estimates. By E. Dobson, 2s.‡

25. *MASONRY AND STONECUTTING:* Rudimentary Treatise on the Principles of Masonic Projection and their application to Construction. By Edward Dobson, M.R.I.B.A., &c. 2s. 6d.‡

42. *COTTAGE BUILDING*. By C. Bruce Allen, Architect. Tenth Edition, revised and enlarged. With a Chapter on Economic Cottages for Allotments, by Edward E. Allen, C.E. 2s.

45. *LIMES, CEMENTS, MORTARS, CONCRETES, MASTICS*, PLASTERING, &c. By G. R. Burnell, C.E. Thirteenth Edition. 1s. 6d.

57. *WARMING AND VENTILATION*. An Exposition of the General Principles as applied to Domestic and Public Buildings, Mines, Lighthouses, Ships, &c. By C. Tomlinson, F.R.S., &c. Illustrated. 3s.

111. *ARCHES, PIERS, BUTTRESSES, &c.:* Experimental Essays on the Principles of Construction. By W. Bland. Illustrated. 1s. 6d.

The ‡ indicates that these vols. may be had strongly bound at 6d. extra.

Architecture, Building, etc., *continued.*

116. *THE ACOUSTICS OF PUBLIC BUILDINGS;* or, The Principles of the Science of Sound applied to the purposes of the Architect and Builder. By T. ROGER SMITH, M.R.I.B.A., Architect. Illustrated. 1s. 6d.

127. *ARCHITECTURAL MODELLING IN PAPER*, the Art of. By T. A. RICHARDSON, Architect. Illustrated. 1s. 6d.

128. *VITRUVIUS — THE ARCHITECTURE OF MARCUS VITRUVIUS POLLO.* In Ten Books. Translated from the Latin by JOSEPH GWILT, F.S.A., F.R.A.S. With 23 Plates. 5s.

130. *GRECIAN ARCHITECTURE*, An Inquiry into the Principles of Beauty in; with an Historical View of the Rise and Progress of the Art in Greece. By the EARL OF ABERDEEN. 1s.

*** *The two preceding Works in One handsome Vol., half bound, entitled "*ANCIENT ARCHITECTURE,*" price 6s.*

132. *THE ERECTION OF DWELLING-HOUSES.* Illustrated by a Perspective View, Plans, Elevations, and Sections of a pair of Semi-detached Villas, with the Specification, Quantities, and Estimates, &c. By S. H. BROOKS. New Edition, with Plates. 2s. 6d.‡

156. *QUANTITIES & MEASUREMENTS* in Bricklayers', Masons', Plasterers', Plumbers', Painters', Paperhangers', Gilders', Smiths', Carpenters' and Joiners' Work. By A. C. BEATON, Surveyor. New Edition. 1s. 6d.

175. *LOCKWOOD'S BUILDER'S PRICE BOOK FOR* 1891. A Comprehensive Handbook of the Latest Prices and Data for Builders, Architects, Engineers, and Contractors. Re-constructed, Re-written, and greatly Enlarged. By FRANCIS T. W. MILLER, A.R.I.B.A. 650 pages. 3s. 6d.; cloth boards, 4s. [*Just Published.*

182. *CARPENTRY AND JOINERY*—THE ELEMENTARY PRIN-CIPLES OF CARPENTRY. Chiefly composed from the Standard Work of THOMAS TREDGOLD, C.E. With a TREATISE ON JOINERY by E. WYNDHAM TARN, M.A. Fifth Edition, Revised. 3s. 6d.‡

182*. *CARPENTRY AND JOINERY. ATLAS* of 35 Plates to accompany the above. With Descriptive Letterpress. 4to. 6s.

185. *THE COMPLETE MEASURER*; the Measurement of Boards, Glass, &c.; Unequal-sided, Square-sided, Octagonal-sided, Round Timber and Stone, and Standing Timber, &c. By RICHARD HORTON. Fifth Edition. 4s.; strongly bound in leather, 5s.

187. *HINTS TO YOUNG ARCHITECTS.* By G. WIGHTWICK. New Edition. By G. H. GUILLAUME. Illustrated. 3s. 6d.‡

188. *HOUSE PAINTING, GRAINING, MARBLING, AND SIGN WRITING:* with a Course of Elementary Drawing for House-Painters, Sign-Writers, &c., and a Collection of Useful Receipts. By ELLIS A. DAVIDSON. Sixth Edition. With Coloured Plates. 5s. cloth limp; 6s. cloth boards.

189. *THE RUDIMENTS OF PRACTICAL BRICKLAYING.* In Six Sections: General Principles; Arch Drawing, Cutting, and Setting; Pointing; Paving, Tiling, Materials; Slating and Plastering; Practical Geometry, Mensuration, &c. By ADAM HAMMOND. Seventh Edition. 1s. 6d.

191. *PLUMBING.* A Text-Book to the Practice of the Art or Craft of the Plumber. With Chapters upon House Drainage and Ventilation. Fifth Edition. With 380 Illustrations. By W. P. BUCHAN. 3s. 6d.‡

192. *THE TIMBER IMPORTER'S, TIMBER MERCHANT'S,* and BUILDER'S STANDARD GUIDE. By R. E. GRANDY. 2s.

206. *A BOOK ON BUILDING*, Civil and *Ecclesiastical*, including CHURCH RESTORATION. With the Theory of Domes and the Great Pyramid, &c. By Sir EDMUND BECKETT, Bart., LL.D., Q.C., F.R.A.S. 4s. 6d.‡

226. *THE JOINTS MADE AND USED BY BUILDERS* in the Construction of various kinds of Engineering and Architectural Works. By WYVILL J. CHRISTY, Architect. With upwards of 160 Engravings on Wood. 3s.‡

228. *THE CONSTRUCTION OF ROOFS OF WOOD AND IRON.* By E. WYNDHAM TARN, M.A., Architect. Second Edition, revised. 1s. 6d.

☞ *The ‡ indicates that these vols. may be had strongly bound at 6d. extra.*

Architecture, Building, etc., *continued.*

229. *ELEMENTARY DECORATION:* as applied to the Interior and Exterior Decoration of Dwelling-Houses, &c. By J. W. FACEY. 2s.

257. *PRACTICAL HOUSE DECORATION.* A Guide to the Art of Ornamental Painting. By JAMES W. FACEY. 2s. 6d.
** *The two preceding Works, in One handsome Vol., half-bound, entitled* "HOUSE DECORATION, ELEMENTARY AND PRACTICAL," *price 5s.*

230. *HANDRAILING.* Showing New and Simple Methods for finding the Pitch of the Plank, Drawing the Moulds, Bevelling, Jointing-up, and Squaring the Wreath. By GEORGE COLLINGS. Second Edition, Revised including A TREATISE ON STAIRBUILDING. Plates and Diagrams. 2s. 6d.

247. *BUILDING ESTATES:* a Rudimentary Treatise on the Development, Sale, Purchase, and General Management of Building Land. By FOWLER MAITLAND, Surveyor. Second Edition, revised. 2s.

248. *PORTLAND CEMENT FOR USERS.* By HENRY FAIJA, Assoc. M. Inst. C.E. Third Edition, corrected. Illustrated. 2s.

252. *BRICKWORK:* a Practical Treatise, embodying the General and Higher Principles of Bricklaying, Cutting and Setting, &c. By F. WALKER. Second Edition, Revised and Enlarged. 1s. 6d.

23. *THE PRACTICAL BRICK AND TILE BOOK.* Comprising:
189. BRICK AND TILE MAKING, by E. DOBSON, A.I.C.E.; PRACTICAL BRICKLAY-
265. ING, by A. HAMMOND; BRICKCUTTING AND SETTING, by A. HAMMOND. 534 pp. with 270 Illustrations. 6s. Strongly half-bound.

253. *THE TIMBER MERCHANT'S, SAW-MILLER'S, AND IMPORTER'S FREIGHT-BOOK AND ASSISTANT.* By WM. RICHARDSON. With a Chapter on Speeds of Saw-Mill Machinery, &c. By M. POWIS BALE, A.M.Inst.C.E. 3s.

258. *CIRCULAR WORK IN CARPENTRY AND JOINERY.* A Practical Treatise on Circular Work of Single and Double Curvature. By GEORGE COLLINGS. Second Edition, 2s. 6d.

259. *GAS FITTING:* A Practical Handbook treating of every Description of Gas Laying and Fitting. By JOHN BLACK. With 122 Illus-. trations. 2s. 6d.‡

261. *SHORING AND ITS APPLICATION:* A Handbook for the Use of Students. By GEORGE H. BLAGROVE. 1s. 6d. [*Just published.*]

265. *THE ART OF PRACTICAL BRICK CUTTING & SETTING.* By ADAM HAMMOND. With 90 Engravings. 1s. 6d. [*Just published.*]

267. *THE SCIENCE OF BUILDING:* An Elementary Treatise on the Principles of Construction. Adapted to the Requirements of Architectural Students. By E. WYNDHAM TARN, M.A. Lond. Third Edition, Revised and Enlarged. With 59 Wood Engravings. 3s. 6d.‡ [*Just published.*]

271. *VENTILATION:* a Text-book to the Practice of the Art of Ventilating Buildings, with a Supplementary Chapter upon Air Testing. By WILLIAM PATON BUCHAN, R.P., Sanitary and Ventilating Engineer, Author of "Plumbing," &c. 3s. 6d.‡ [*Just published.*]

SHIPBUILDING, NAVIGATION, MARINE ENGINEERING, ETC.

51. *NAVAL ARCHITECTURE.* An Exposition of the Elementary Principles of the Science, and their Practical Application to Naval Construction. By J. PEAKE. Fifth Edition, with Plates and Diagrams. 3s. 6d.‡

53*. *SHIPS FOR OCEAN & RIVER SERVICE,* Elementary and Practical Principles of the Construction of. By H. A. SOMMERFELDT. 1s. 6d.

53**. *AN ATLAS OF ENGRAVINGS* to Illustrate the above. Twelve large folding plates. Royal 4to, cloth. 7s. 6d.

54. *MASTING, MAST-MAKING, AND RIGGING OF SHIPS,* Also Tables of Spars, Rigging, Blocks; Chain, Wire, and Hemp Ropes, &c., relative to every class of vessels. By ROBERT KIPPING, N.A. 2s.

☞ *The ‡ indicates that these vols. may be had strongly bound at 6d. extra.*

LONDON : CROSBY LOCKWOOD AND SON,

Shipbuilding, Navigation, Marine Engineering, etc., *cont.*

54*. *IRON SHIP-BUILDING.* With Practical Examples and Details.
By JOHN GRANTHAM, C.E. Fifth Edition. 4s.

55. *THE SAILOR'S SEA BOOK:* a Rudimentary Treatise on Navigation. By JAMES GREENWOOD, B.A. With numerous Woodcuts and Coloured Plates. New and enlarged edition. By W. H. ROSSER. 2s. 6d.‡

80. *MARINE ENGINES AND STEAM VESSELS.* By ROBERT MURRAY, C.E. Eighth Edition, thoroughly Revised, with Additions by the Author and by GEORGE CARLISLE, C.E. 4s. 6d. limp; 5s. cloth boards.

83*bis.* *THE FORMS OF SHIPS AND BOATS.* By W. BLAND. Seventh Edition, Revised, with numerous Illustrations and Models. 1s. 6d.

99. *NAVIGATION AND NAUTICAL ASTRONOMY*, in Theory and Practice. By Prof. J. R. YOUNG. New Edition. 2s. 6d.

106. *SHIPS' ANCHORS*, a Treatise on. By G. COTSELL, N.A. 1s. 6d.

149. *SAILS AND SAIL-MAKING.* With Draughting, and the Centre of Effort of the Sails; Weights and Sizes of Ropes; Masting, Rigging, and Sails of Steam Vessels, &c. 12th Edition. By R. KIPPING, N.A., 2s. 6d.‡

155. *ENGINEER'S GUIDE TO THE ROYAL & MERCANTILE* NAVIES. By a PRACTICAL ENGINEER. Revised by D. F. M'CARTHY. 3s.

55 *PRACTICAL NAVIGATION.* Consisting of The Sailor's
& Sea-Book. By JAMES GREENWOOD and W. H. ROSSER. Together with
204. the requisite Mathematical and Nautical Tables for the Working of the Problems. By H. LAW, C.E., and Prof. J. R. YOUNG. 7s. Half-bound.

AGRICULTURE, GARDENING, ETC.

61*. *A COMPLETE READY RECKONER FOR THE ADMEA-* SUREMENT OF LAND, &c. By A. ARMAN. Third Edition, revised and extended by C. NORRIS, Surveyor, Valuer, &c. 2s.

131. *MILLER'S, CORN MERCHANT'S, AND FARMER'S* READY RECKONER. Second Edition, with a Price List of Modern Flour-Mill Machinery, by W. S. HUTTON, C.E. 2s.

140. *SOILS, MANURES, AND CROPS.* (Vol. 1. OUTLINES OF MODERN FARMING.) By R. SCOTT BURN. Woodcuts. 2s.

141. *FARMING & FARMING ECONOMY*, Notes, Historical and Practical, on. (Vol. 2. OUTLINES OF MODERN FARMING.) By R. SCOTT BURN. 3s.

142. *STOCK; CATTLE, SHEEP, AND HORSES.* (Vol. 3. OUTLINES OF MODERN FARMING.) By R. SCOTT BURN. Woodcuts. 2s. 6d.

145. *DAIRY, PIGS, AND POULTRY*, Management of the. By R. SCOTT BURN. (Vol. 4. OUTLINES OF MODERN FARMING.) 2s.

146. *UTILIZATION OF SEWAGE, IRRIGATION, AND* RECLAMATION OF WASTE LAND. (Vol. 5. OUTLINES OF MODERN FARMING.) By R. SCOTT BURN. Woodcuts. 2s. 6d.

⁎ *Nos.* 140-1-2-5-6, *in One Vol., handsomely half-bound, entitled* " OUTLINES OF MODERN FARMING." *By* ROBERT SCOTT BURN. *Price* 12s.

177. *FRUIT TREES*, The Scientific and Profitable Culture of. From the French of DU BREUIL. Revised by GEO. GLENNY. 187 Woodcuts. 3s. 6d.‡

198. *SHEEP; THE HISTORY, STRUCTURE, ECONOMY, AND* DISEASES OF. By W. C. SPOONER, M.R.V.C., &c. Fifth Edition, enlarged, including Specimens ot New and Improved Breeds. 3s. 6d.‡

201. *KITCHEN GARDENING MADE EASY.* By GEORGE M. F. GLENNY. Illustrated. 1s. 6d.‡

207. *OUTLINES OF FARM MANAGEMENT, and the Organi-* zation of Farm Labour. By R. SCOTT BURN. 2s. 6d.‡

208. *OUTLINES OF LANDED ESTATES MANAGEMENT.* By R. SCOTT BURN. 2s. 6d.

⁎ *Nos.* 207 & 208 *in One Vol., handsomely half-bound, entitled* " OUTLINES OF LANDED ESTATES AND FARM MANAGEMENT." *By* R. SCOTT BURN. *Price* 6s.

☞ *The* ‡ *indicates that these vols. may be had strongly bound at 6d. extra.*

Agriculture, Gardening, etc., *continued.*

209. *THE TREE PLANTER AND PLANT PROPAGATOR.*
A Practical Manual on the Propagation of Forest Trees, Fruit Trees,
Flowering Shrubs, Flowering Plants, &c. By SAMUEL WOOD. 2s.

210. *THE TREE PRUNER.* A Practical Manual on the Pruning of
Fruit Trees, including also their Training and Renovation; also the Pruning
of Shrubs, Climbers, and Flowering Plants. By SAMUEL WOOD. 1s. 6d.

⁎ *Nos.* 209 & 210 *in One Vol., handsomely half-bound, entitled* "THE TREE
PLANTER, PROPAGATOR, AND PRUNER." By SAMUEL WOOD. *Price* 3s. 6d.

218. *THE HAY AND STRAW MEASURER :* Being New Tables
for the Use of Auctioneers, Valuers, Farmers, Hay and Straw Dealers, &c.
By JOHN STEELE. Fourth Edition. 2s.

222. *SUBURBAN FARMING.* The Laying-out and Cultivation of
Farms, adapted to the Produce of Milk, Butter, and Cheese, Eggs, Poultry,
and Pigs. By Prof. JOHN DONALDSON and R. SCOTT BURN. 3s. 6d.‡

231. *THE ART OF GRAFTING AND BUDDING.* By CHARLES
BALTET. With Illustrations. 2s. 6d.‡

232. *COTTAGE GARDENING;* or, Flowers, Fruits, and Vegetables
for Small Gardens. By E. HOBDAY. 1s. 6d.

233. *GARDEN RECEIPTS.* Edited by CHARLES W. QUIN. 1s. 6d.

234. *MARKET AND KITCHEN GARDENING.* By C. W. SHAW,
'late Editor of "Gardening Illustrated." 3s.‡   [*Just published.*

239. *DRAINING AND EMBANKING.* A Practical Treatise, em-
bodying the most recent experience in the Application of Improved Methods.
By JOHN SCOTT, late Professor of Agriculture and Rural Economy at the
Royal Agricultural College, Cirencester. With 68 Illustrations. 1s. 6d.

240. *IRRIGATION AND WATER SUPPLY.* A Treatise on Water
Meadows, Sewage Irrigation, and Warping; the Construction of Wells,
Ponds, and Reservoirs, &c. By Prof. JOHN SCOTT. With 34 Illus. 1s. 6d.

241. *FARM ROADS, FENCES, AND GATES.* A Practical
Treatise on the Roads, Tramways, and Waterways of the Farm; the
Principles of Enclosures; and the different kinds of Fences, Gates, and
Stiles. By Professor JOHN SCOTT. With 75 Illustrations. 1s. 6d.

242. *FARM BUILDINGS.* A Practical Treatise on the Buildings
necessary for various kinds of Farms, their Arrangement and Construction,
with Plans and Estimates. By Prof. JOHN SCOTT. With 105 Illus. 2s.

243. *BARN IMPLEMENTS AND MACHINES.* A Practical
Treatise on the Application of Power to the Operations of Agriculture; and
on various Machines used in the Threshing-barn, in the Stock-yard, and in the
Dairy, &c. By Prof. J. SCOTT. With 123 Illustrations. 2s.

244. *FIELD IMPLEMENTS AND MACHINES.* A Practical
Treatise on the Varieties now in use, with Principles and Details of Con-
struction, their Points of Excellence, and Management. By Professor JOHN
SCOTT. With 138 Illustrations. 2s.

245. *AGRICULTURAL SURVEYING.* A Practical Treatise on
Land Surveying, Levelling, and Setting-out; and on Measuring and Esti-
mating Quantities, Weights, and Values of Materials, Produce, Stock, &c.
By Prof. JOHN SCOTT. With 62 Illustrations. 1s. 6d.

⁎ *Nos.* 239 *to* 245 *in One Vol., handsomely half-bound, entitled* "THE COMPLETE
TEXT-BOOK OF FARM ENGINEERING." By Professor JOHN SCOTT. *Price* 12s.

250. *MEAT PRODUCTION.* A Manual for Producers, Distributors,
&c. By JOHN EWART. 2s. 6d.‡

266. *BOOK-KEEPING FOR FARMERS & ESTATE OWNERS.*
By J. M. WOODMAN, Chartered Accountant. 2s. 6d. cloth limp; 3s. 6d.
cloth boards.   [*Just published.*

☞ *The* ‡ *indicates that these vols. may be had strongly bound at* 6d. *extra.*

MATHEMATICS, ARITHMETIC, ETC.

32. *MATHEMATICAL INSTRUMENTS*, a Treatise on; Their Construction, Adjustment, Testing, and Use concisely Explained. By J. F. HEATHER, M.A. Fourteenth Edition, revised, with additions, by A. T. WALMISLEY, M.I.C.E., Fellow of the Surveyors' Institution. Original Edition, in 1 vol., Illustrated. 2s.‡ [*Just published.*

. In ordering the above, be careful to say, " Original Edition " (No. 32), to distinguish it from the Enlarged Edition in 3 vols. (Nos. 168-9-70.)

76. *DESCRIPTIVE GEOMETRY*, an Elementary Treatise on; with a Theory of Shadows and of Perspective, extracted from the French of G. MONGE. To which is added, a description of the Principles and Practice of Isometrical Projection. By J. F. HEATHER, M.A. With 14 Plates. 2s.

178. *PRACTICAL PLANE GEOMETRY:* giving the Simplest Modes of Constructing Figures contained in one Plane and Geometrical Construction of the Ground. By J. F. HEATHER, M.A. With 215 Woodcuts. 2s.

83. *COMMERCIAL BOOK-KEEPING*. With Commercial Phrases and Forms in English, French, Italian, and German. By JAMES HADDON, M.A., Arithmetical Master of King's College School, London. 1s. 6d.

84. *ARITHMETIC*, a Rudimentary Treatise on: with full Explanations of its Theoretical Principles, and numerous Examples for Practice. By Professor J. R. YOUNG. Eleventh Edition. 1s. 6d.

84*. A KEY to the above, containing Solutions in full to the Exercises, together with Comments, Explanations, and Improved Processes, for the Use of Teachers and Unassisted Learners. By J. R. YOUNG. 1s. 6d.

85. *EQUATIONAL ARITHMETIC*, applied to Questions of Interest, Annuities, Life Assurance, and General Commerce ; with various Tables by which all Calculations may be greatly facilitated. By W. HIPSLEY. 2s.

86. *ALGEBRA*, the Elements of. By JAMES HADDON, M.A. With Appendix, containing miscellaneous Investigations, and a Collection of Problems in various parts of Algebra. 2s.

86*. A KEY AND COMPANION to the above Book, forming an extensive repository of Solved Examples and Problems in Illustration of the various Expedients necessary in Algebraical Operations. By J. R. YOUNG. 1s. 6d.

88. *EUCLID*, THE ELEMENTS OF : with many additional Propositions
89. and Explanatory Notes: to which is prefixed, an Introductory Essay on Logic. By HENRY LAW, C.E. 2s. 6d.‡

. Sold also separately, viz. :—

88. EUCLID, The First Three Books. By HENRY LAW, C.E. 1s. 6d.
89. EUCLID, Books 4, 5, 6, 11, 12. By HENRY LAW, C.E. 1s. 6d.

90. *ANALYTICAL GEOMETRY AND CONIC SECTIONS*, By JAMES HANN. A New Edition, by Professor J. R. YOUNG. 2s.‡

91. *PLANE TRIGONOMETRY*, the Elements of. By JAMES HANN, formerly Mathematical Master of King's College, London. 1s. 6d.

92. *SPHERICAL TRIGONOMETRY*, the Elements of. By JAMES HANN. Revised by CHARLES H. DOWLING, C.E. 1s.

. Or with " The Elements of Plane Trigonometry," in One Volume, 2s. 6d.

93. *MENSURATION AND MEASURING*. With the Mensuration and Levelling of Land for the Purposes of Modern Engineering. By T. BAKER, C.E. New Edition by E. NUGENT, C.E. Illustrated. 1s. 6d.

101. *DIFFERENTIAL CALCULUS*, Elements of the. By W. S. B. WOOLHOUSE, F.R.A.S., &c. 1s. 6d.

102. *INTEGRAL CALCULUS*, Rudimentary Treatise on the. By HOMERSHAM COX, B.A. Illustrated. 1s.

136. *ARITHMETIC*, Rudimentary, for the Use of Schools and Self-Instruction. By JAMES HADDON, M.A. Revised by A. ARMAN. 1s. 6d.

137. A KEY TO HADDON'S RUDIMENTARY ARITHMETIC. By A. ARMAN. 1s. 6d.

The ‡ indicates that these vols. may be had strongly bound at 6d. extra.

Mathematics, Arithmetic, etc., *continued.*

168. *DRAWING AND MEASURING INSTRUMENTS.* Including—I. Instruments employed in Geometrical and Mechanical Drawing, and in the Construction, Copying, and Measurement of Maps and Plans. II. Instruments used for the purposes of Accurate Measurement, and for Arithmetical Computations. By J. F. HEATHER, M.A. Illustrated. 1s. 6d.

169. *OPTICAL INSTRUMENTS.* Including (more especially) Telescopes, Microscopes, and Apparatus for producing copies of Maps and Plans by Photography. By J. F. HEATHER, M.A. Illustrated. 1s. 6d.

170. *SURVEYING AND ASTRONOMICAL INSTRUMENTS.* Including—I. Instruments Used for Determining the Geometrical Features of a portion of Ground. II. Instruments Employed in Astronomical Observations. By J. F. HEATHER, M.A. Illustrated. 1s. 6d.

⁎⁎ The above three volumes form an enlargement of the Author's original work "Mathematical Instruments." (See No. 32 in the Series.)

168.⎫
169.⎬ *MATHEMATICAL INSTRUMENTS.* By J. F. HEATHER, M.A. Enlarged Edition, for the most part entirely re-written. The 3 Parts as
170.⎭ above, in One thick Volume. With numerous Illustrations. 4s. 6d.‡

158. *THE SLIDE RULE, AND HOW TO USE IT;* containing full, easy, and simple Instructions to perform all Business Calculations with unexampled rapidity and accuracy. By CHARLES HOARE, C.E. Fifth Edition. With a Slide Rule in tuck of cover. 2s. 6d.‡

196. *THEORY OF COMPOUND INTEREST AND ANNUITIES;* with Tables of Logarithms for the more Difficult Computations of Interest, Discount, Annuities, &c. By FÉDOR THOMAN. 4s.‡

199. *THE COMPENDIOUS CALCULATOR;* or, Easy and Concise Methods of Performing the various Arithmetical Operations required in Commercial and Business Transactions; together with Useful Tables. By D. O'GORMAN. Twenty-seventh Edition, carefully revised by C. NORRIS. 2s. 6d., cloth limp; 3s. 6d., strongly half-bound in leather.

204. *MATHEMATICAL TABLES,* for Trigonometrical, Astronomical, and Nautical Calculations; to which is prefixed a Treatise on Logarithms. By HENRY LAW, C.E. Together with a Series of Tables for Navigation and Nautical Astronomy. By Prof. J. R. YOUNG. New Edition. 4s.

204*. *LOGARITHMS.* With Mathematical Tables for Trigonometrical, Astronomical, and Nautical Calculations. By HENRY LAW, M.Inst.C.E. New and Revised Edition. (Forming part of the above Work). 3s.

221. *MEASURES, WEIGHTS, AND MONEYS OF ALL NATIONS,* and an Analysis of the Christian, Hebrew, and Mahometan Calendars. By W. S. B. WOOLHOUSE, F.R.A.S., F.S.S. Seventh Edition. 2s. 6d.‡

227. *MATHEMATICS AS APPLIED TO THE CONSTRUCTIVE ARTS.* Illustrating the various processes of Mathematical Investigation, by means of Arithmetical and Simple Algebraical Equations and Practical Examples. By FRANCIS CAMPIN. C.E. Second Edition. 3s.‡

PHYSICAL SCIENCE, NATURAL PHILOSOPHY, ETC.

1. *CHEMISTRY.* By Professor GEORGE FOWNES, F.R.S. With an Appendix on the Application of Chemistry to Agriculture. 1s.

2. *NATURAL PHILOSOPHY,* Introduction to the Study of. By C. TOMLINSON. Woodcuts. 1s. 6d.

6. *MECHANICS,* Rudimentary Treatise on. By CHARLES TOMLINSON. Illustrated. 1s. 6d.

7. *ELECTRICITY;* showing the General Principles of Electrical Science, and the purposes to which it has been applied. By Sir W. SNOW HARRIS, F.R.S., &c. With Additions by R. SABINE, C.E., F.S.A. 1s. 6d.

7*. *GALVANISM.* By Sir W. SNOW HARRIS. New Edition by ROBERT SABINE, C.E., F.S.A. 1s. 6d.

8. *MAGNETISM;* being a concise Exposition of the General Principles of Magnetical Science. By Sir W. SNOW HARRIS. New Edition, revised by H. M. NOAD, Ph.D. With 165 Woodcuts. 3s. 6d.‡

The ‡ indicates that these vols. may be had strongly bound at 6d. extra.

LONDON : CROSBY LOCKWOOD AND SON,

Physical Science, Natural Philosophy, etc., *continued*.

11. *THE ELECTRIC TELEGRAPH;* its History and Progress; with Descriptions of some of the Apparatus. By R. SABINE, C.E., F.S.A. 3s.

12. *PNEUMATICS*, including Acoustics and the Phenomena of Wind Currents, for the Use of Beginners. By CHARLES TOMLINSON, F.R.S. Fourth Edition, enlarged. Illustrated. 1s. 6d. [*Just published.*

72. *MANUAL OF THE MOLLUSCA;* a Treatise on Recent and Fossil Shells. By Dr. S. P. WOODWARD, A.L.S. Fourth Edition. With Plates and 300 Woodcuts. 7s. 6d., cloth.

96. *ASTRONOMY.* By the late Rev. ROBERT MAIN, M.A. Third Edition, by WILLIAM THYNNE LYNN, B.A., F.R.A.S. 2s.

97. *STATICS AND DYNAMICS*, the Principles and Practice of; embracing also a clear development of Hydrostatics, Hydrodynamics, and Central Forces. By T. BAKER, C.E. Fourth Edition. 1s. 6d.

173. *PHYSICAL GEOLOGY*, partly based on Major-General PORT-LOCK's "Rudiments of Geology." By RALPH TATE, A.L.S., &c. Woodcuts. 2s.

174. *HISTORICAL GEOLOGY*, partly based on Major-General PORTLOCK's "Rudiments." By RALPH TATE, A.L.S., &c. Woodcuts. 2s. 6d.

173 *RUDIMENTARY TREATISE ON GEOLOGY*, Physical and
& Historical. Partly based on Major-General PORTLOCK's "Rudiments of
174. Geology." By RALPH TATE, A.L.S., F.G.S., &c. In One Volume. 4s. 6d.‡

183 *ANIMAL PHYSICS*, Handbook of. By Dr. LARDNER,
& formerly Professor of Natural Philosophy and Astronomy in University
184. College, Lond. With 520 Illustrations. In One Vol. 7s. 6d., cloth boards.
⁎ *Sold also in Two Parts, as follows :—*
183. ANIMAL PHYSICS. By Dr. LARDNER. Part I., Chapters I.—VII. 4s.
184. ANIMAL PHYSICS. By Dr. LARDNER. Part II., Chapters VIII.—XVIII. 3s.

269. *LIGHT :* an Introduction to the Science of Optics, for the Use of Students of Architecture, Engineering, and other Applied Sciences. By E. WYNDHAM TARN, M.A. 1s. 6d. ⁚ [*Just published.*

FINE ARTS.

20. *PERSPECTIVE FOR BEGINNERS.* Adapted to Young Students and Amateurs in Architecture, Painting, &c. By GEORGE PYNE. 2s.

40 *GLASS STAINING, AND THE ART OF PAINTING ON GLASS.* From the German of Dr. GESSERT and EMANUEL OTTO FROM-BERG. With an Appendix on THE ART OF ENAMELLING. 2s. 6d.

69. *MUSIC*, A Rudimentary and Practical Treatise on. With numerous Examples. By CHARLES CHILD SPENCER. 2s. 6d.

71. *PIANOFORTE*, The Art of Playing the. With numerous Exer-cises & Lessons from the Best Masters. By CHARLES CHILD SPENCER. 1s.6d.

69-71. *MUSIC & THE PIANOFORTE.* In one vol. Half bound, 5s.

181. *PAINTING POPULARLY EXPLAINED*, including Fresco, Oil, Mosaic, Water Colour, Water-Glass, Tempera, Encaustic, Miniature, Painting on Ivory, Vellum, Pottery, Enamel, Glass, &c. With Historical Sketches of the Progress of the Art by THOMAS JOHN GULLICK, assisted by JOHN TIMBS, F.S.A. Fifth Edition, revised and enlarged. 5s.‡

186. *A GRAMMAR OF COLOURING*, applied to Decorative Painting and the Arts. By GEORGE FIELD. New Edition, enlarged and adapted to the Use of the Ornamental Painter and Designer. By ELLIS A. DAVIDSON. With two new Coloured Diagrams, &c. 3s.‡

246. *A DICTIONARY OF PAINTERS, AND HANDBOOK FOR* PICTURE AMATEURS; including Methods of Painting, Cleaning, Re-lining and Restoring, Schools of Painting, &c. With Notes on the Copyists and Imitators of each Master. By PHILIPPE DARYL. 2s. 6d.‡

☞ *The ‡ indicates that these vols. may be had strongly bound at 6d. extra.*

INDUSTRIAL AND USEFUL ARTS.

23. *BRICKS AND TILES*, Rudimentary Treatise on the Manufacture of. By E. DOBSON, M.R.I.B.A. Illustrated, 3s.‡

67. *CLOCKS, WATCHES, AND BELLS*, a Rudimentary Treatise on. By Sir EDMUND BECKETT, LL.D., Q.C. Seventh Edition, revised and enlarged. 4s. 6d. limp ; 5s. 6d. cloth boards.

83**. *CONSTRUCTION OF DOOR LOCKS.* Compiled from the Papers of A. C. HOBBS, and Edited by CHARLES TOMLINSON. F.R.S. 2s. 6d.

162. *THE BRASS FOUNDER'S MANUAL ;* Instructions for Modelling, Pattern-Making, Moulding, Turning, Filing, Burnishing, Bronzing, &c. With copious Receipts, &c. By WALTER GRAHAM. 2s.‡

205. *THE ART OF LETTER PAINTING MADE EASY.* By J. G. BADENOCH. Illustrated with 12 full-page Engravings of Examples. 1s. 6d.

215. *THE GOLDSMITH'S HANDBOOK,* containing full Instructions for the Alloying and Working of Gold. By GEORGE E. GEE, 3s.‡

225. *THE SILVERSMITH'S HANDBOOK,* containing full Instructions for the Alloying and Working of Silver. By GEORGE E. GEE. 3s.‡
** *The two preceding Works, in One handsome Vol., half-bound, entitled* "THE GOLDSMITH'S & SILVERSMITH'S COMPLETE HANDBOOK," 7s.

249. *THE HALL-MARKING OF JEWELLERY PRACTICALLY CONSIDERED.* By GEORGE E. GEE. 3s.‡

224. *COACH BUILDING,* A Practical Treatise, Historical and Descriptive. By J. W. BURGESS. 2s. 6d.‡

235. *PRACTICAL ORGAN BUILDING.* By W. E. DICKSON, M.A., Precentor of Ely Cathedral. Illustrated. 2s. 6d.‡

262. *THE ART OF BOOT AND SHOEMAKING.* By JOHN BEDFORD LENO. Numerous Illustrations. Third Edition. 2s.

263. *MECHANICAL DENTISTRY :* A Practical Treatise on the Construction of the Various Kinds of Artificial Dentures, with Formulæ, Tables, Receipts, &c. By CHARLES HUNTER. Third Edition. 3s.‡

270. *WOOD ENGRAVING :* A Practical and Easy Introduction to the Study of the Art. By W. N. BROWN. 1s. 6d.

MISCELLANEOUS VOLUMES.

36. *A DICTIONARY OF TERMS used in ARCHITECTURE, BUILDING, ENGINEERING, MINING, METALLURGY, ARCHÆOLOGY, the FINE ARTS, &c.* By JOHN WEALE. Fifth Edition. Revised by ROBERT HUNT, F.R.S. Illustrated. 5s. limp ; 6s. cloth boards.

50. *THE LAW OF CONTRACTS FOR WORKS AND SERVICES.* By DAVID GIBBONS. Third Edition, enlarged. 3s.‡

112. *MANUAL OF DOMESTIC MEDICINE.* By R. GOODING, B.A., M.D. A Family Guide in all Cases of Accident and Emergency. 2s.

112*. *MANAGEMENT OF HEALTH.* A Manual of Home and Personal Hygiene. By the Rev. JAMES BAIRD, B.A. 1s.

150. *LOGIC,* Pure and Applied. By S. H. EMMENS. 1s. 6d.

153. *SELECTIONS FROM LOCKE'S ESSAYS ON THE HUMAN UNDERSTANDING.* With Notes by S. H. EMMENS. 2s.

154. *GENERAL HINTS TO EMIGRANTS.* 2s.

157. *THE EMIGRANT'S GUIDE TO NATAL.* By ROBERT JAMES MANN, F.R.A.S., F.M.S. Second Edition. Map. 2s.

193. *HANDBOOK OF FIELD FORTIFICATION.* By Major W. W. KNOLLYS, F.R.G.S. With 163 Woodcuts. 3s.‡

194. *THE HOUSE MANAGER :* Being a Guide to Housekeeping. Practical Cookery, Pickling and Preserving, Household Work, Dairy Management, &c. By AN OLD HOUSEKEEPER. 3s. 6d.‡

194, *HOUSE BOOK (The).* Comprising :—I. THE HOUSE MANAGER.
112 & By an OLD HOUSEKEEPER. II. DOMESTIC MEDICINE. By R. GOODING, M.D.
112*. III. MANAGEMENT OF HEALTH. By J. BAIRD. In One Vol., half-bound, 6s.

☞ The ‡ indicates that these vols. may be had strongly bound at 6d. extra.

LONDON : CROSBY LOCKWOOD AND SON,

EDUCATIONAL AND CLASSICAL SERIES.

HISTORY.

1. **England, Outlines of the History of**; more especially with reference to the Origin and Progress of the English Constitution. By WILLIAM DOUGLAS HAMILTON, F.S.A., of Her Majesty's Public Record Office. 4th Edition, revised. 5s.; cloth boards, 6s.

5. **Greece, Outlines of the History of**; in connection with the Rise of the Arts and Civilization in Europe. By W. DOUGLAS HAMILTON, of University College, London, and EDWARD LEVIEN, M.A., of Balliol College, Oxford. 2s. 6d.; cloth boards, 3s. 6d.

7. **Rome, Outlines of the History of**; from the Earliest Period to the Christian Era and the Commencement of the Decline of the Empire. By EDWARD LEVIEN, of Balliol College, Oxford. Map, 2s. 6d.; cl. bds. 3s. 6d.

9. **Chronology of History, Art, Literature, and Progress,** from the Creation of the World to the Present Time. The Continuation by W. D. HAMILTON, F.S.A. 3s.; cloth boards, 3s. 6d.

50. **Dates and Events in English History,** for the use of Candidates in Public and Private Examinations. By the Rev. E. RAND. 1s.

ENGLISH LANGUAGE AND MISCELLANEOUS.

11. **Grammar of the English Tongue,** Spoken and Written. With an Introduction to the Study of Comparative Philology. By HYDE CLARKE, D.C.L. Fourth Edition. 1s. 6d.

12. **Dictionary of the English Language,** as Spoken and Written. Containing above 100,000 Words. By HYDE CLARKE, D.C.I. 3s. 6d.; cloth boards, 4s. 6d.; complete with the GRAMMAR, cloth bds., 5s. 6d.

48. **Composition and Punctuation,** familiarly Explained for those who have neglected the Study of Grammar. By JUSTIN BRENAN. 18th Edition. 1s. 6d.

49. **Derivative Spelling-Book:** Giving the Origin of Every Word from the Greek, Latin, Saxon, German, Teutonic, Dutch, French, Spanish, and other Languages; with their present Acceptation and Pronunciation. By J. ROWBOTHAM, F.R.A.S. Improved Edition. 1s. 6d.

51. **The Art of Extempore Speaking:** Hints for the Pulpit, the Senate, and the Bar. By M. BAUTAIN, Vicar-General and Professor at the Sorbonne. Translated from the French. 8th Edition, carefully corrected. 2s. 6d.

54. **Analytical Chemistry,** Qualitative and Quantitative, a Course of. To which is prefixed, a Brief Treatise upon Modern Chemical Nomenclature and Notation. By WM. W. PINK and GEORGE E. WEBSTER. 2s.

THE SCHOOL MANAGERS' SERIES OF READING BOOKS,

Edited by the Rev. A. R. GRANT, Rector of Hitcham, and Honorary Canon of Ely; formerly H.M. Inspector of Schools. INTRODUCTORY PRIMER, 3d.

	s.	d.							s.	d.
FIRST STANDARD	. 0	6	FOURTH STANDARD	.	.	.	.	.	1	2
SECOND „	. 0	10	FIFTH „	.	.	.	.	.	1	6
THIRD „	. 1	0	SIXTH „	.	.	.	.	.	1	6

LESSONS FROM THE BIBLE. Part I. Old Testament. 1s.
LESSONS FROM THE BIBLE. Part II. New Testament, to which is added THE GEOGRAPHY OF THE BIBLE, for very young Children. By Rev. C. THORNTON FORSTER. 1s. 2d. *₊* Or the Two Parts in One Volume. 2s.

FRENCH.

24. **French Grammar.** With Complete and Concise Rules on the Genders of French Nouns. By G. L. STRAUSS, Ph.D. 1s. 6d.
25. **French–English Dictionary.** Comprising a large number of New Terms used in Engineering, Mining, &c. By ALFRED ELWES. 1s. 6d.
26. **English–French Dictionary.** By ALFRED ELWES. 2s.
25,26. **French Dictionary** (as above). Complete, in One Vol., 3s.; cloth boards, 3s. 6d. *** Or with the GRAMMAR, cloth boards, 4s. 6d.
47. **French and English Phrase Book :** containing Introductory Lessons, with Translations, several Vocabularies of Words, a Collection of suitable Phrases, and Easy Familiar Dialogues. 1s. 6d.

GERMAN.

39. **German Grammar.** Adapted for English Students, from Heyse's Theoretical and Practical Grammar, by Dr. G. L. STRAUSS. 1s. 6d.
40. **German Reader :** A Series of Extracts, carefully culled from the most approved Authors of Germany; with Notes, Philological and Explanatory. By G. L. STRAUSS, Ph.D. 1s.
41-43. **German Triglot Dictionary.** By N. E. S. A. HAMILTON. In Three Parts. Part I. German-French-English. Part II. English-German-French. Part III. French-German-English. 3s., or cloth boards, 4s.
41-43 **German Triglot Dictionary** (as above), together with German
& 39. Grammar (No. 39), in One Volume, cloth boards, 5s.

ITALIAN.

27. **Italian Grammar,** arranged in Twenty Lessons, with a Course of Exercises. By ALFRED ELWES. 1s. 6d.
28. **Italian Triglot Dictionary,** wherein the Genders of all the Italian and French Nouns are carefully noted down. By ALFRED ELWES. Vol. 1. Italian-English-French. 2s. 6d.
30. **Italian Triglot Dictionary.** By A. ELWES. Vol. 2. English-French-Italian. 2s. 6d.
32. **Italian Triglot Dictionary.** By ALFRED ELWES. Vol. 3. French-Italian-English. 2s. 6d.
28,30, **Italian Triglot Dictionary** (as above). In One Vol., 7s. 6d.
32. Cloth boards.

SPANISH AND PORTUGUESE.

34. **Spanish Grammar,** in a Simple and Practical Form. With a Course of Exercises. By ALFRED ELWES. 1s. 6d.
35. **Spanish–English and English–Spanish Dictionary.** Including a large number of Technical Terms used in Mining, Engineering, &c. with the proper Accents and the Gender of every Noun. By ALFRED ELWES 4s. ; cloth boards, 5s. *** Or with the GRAMMAR, cloth boards, 6s.
55. **Portuguese Grammar,** in a Simple and Practical Form. With a Course of Exercises. By ALFRED ELWES. 1s. 6d.
56. **Portuguese–English and English–Portuguese Dictionary.** Including a large number of Technical Terms used in Mining, Engineering, &c., with the proper Accents and the Gender of every Noun. By ALFRED ELWES. Second Edition, Revised, 5s. ; cloth boards, 6s. *** Or with the GRAMMAR, cloth boards, 7s.

HEBREW.

46*. **Hebrew Grammar.** By Dr. BRESSLAU. 1s. 6d.
44. **Hebrew and English Dictionary,** Biblical and Rabbinical; containing the Hebrew and Chaldee Roots of the Old Testament Post-Rabbinical Writings. By Dr. BRESSLAU. 6s.
46. **English and Hebrew Dictionary.** By Dr. BRESSLAU. 3s.
44,46. **Hebrew Dictionary** (as above), in Two Vols., complete, with
46*. the GRAMMAR, cloth boards, 12s.

LATIN.

19. **Latin Grammar.** Containing the Inflections and Elementary Principles of Translation and Construction. By the Rev. THOMAS GOODWIN, M.A., Head Master of the Greenwich Proprietary School. 1s. 6d.

20. **Latin-English Dictionary.** By the Rev. THOMAS GOODWIN, M.A. 2s.

22. **English-Latin Dictionary;** together with an Appendix of French and Italian Words which have their origin from the Latin. By the Rev. THOMAS GOODWIN, M.A. 1s. 6d.

20,22. **Latin Dictionary** (as above). Complete in One Vol., 3s. 6d. cloth boards, 4s. 6d. °°° Or with the GRAMMAR, cloth boards, 5s. 6d.

LATIN CLASSICS. With Explanatory Notes in English.

1. **Latin Delectus.** Containing Extracts from Classical Authors, with Genealogical Vocabularies and Explanatory Notes, by H. YOUNG. 1s. 6d.

2. **Cæsaris Commentarii de Bello Gallico.** Notes, and a Geographical Register for the Use of Schools, by H. YOUNG. 2s.

3. **Cornelius Nepos.** With Notes. By H. YOUNG. 1s.

4. **Virgilii** Maronis Bucolica et Georgica. With Notes on the Bucolics by W. RUSHTON, M.A., and on the Georgics by H. YOUNG. 1s. 6d.

5. **Virgilii** Maronis Æneis. With Notes, Critical and Explanatory, by H. YOUNG. New Edition, revised and improved. With copious Additional Notes by Rev. T. H. L. LEARY, D.C.L., formerly Scholar of Brasenose College, Oxford. 3s.

5* ———— Part 1. Books i.—vi., 1s. 6d.
5** ———— Part 2. Books vii.—xii., 2s.

6. **Horace;** Odes, Epode, and Carmen Sæculare. Notes by H. YOUNG. 1s. 6d.

7. **Horace;** Satires, Epistles, and Ars Poetica. Notes by W. BROWNRIGG SMITH, M.A., F.R.G.S. 1s. 6d.

8. **Sallustii** Crispi Catalina et Bellum Jugurthinum. Notes, Critical and Explanatory, by W. M. DONNE, B.A., Trin. Coll., Cam. 1s. 6d.

9. **Terentii** Andria et Heautontimorumenos. With Notes, Critical and Explanatory, by the Rev. JAMES DAVIES, M.A. 1s. 6d.

10. **Terentii** Adelphi, Hecyra, Phormio. Edited, with Notes, Critical and Explanatory, by the Rev. JAMES DAVIES, M.A. 2s.

11. **Terentii** Eunuchus, Comœdia. Notes, by Rev. J. DAVIES, M.A. 1s. 6d.

12. **Ciceronis** Oratio pro Sexto Roscio Amerino. Edited, with an Introduction, Analysis, and Notes, Explanatory and Critical, by the Rev. JAMES DAVIES, M.A. 1s. 6d.

13. **Ciceronis** Orationes in Catilinam, Verrem, et pro Archia. With Introduction, Analysis, and Notes, Explanatory and Critical, by Rev. T. H. L. LEARY, D.C.L. formerly Scholar of Brasenose College, Oxford. 1s. 6d.

14. **Ciceronis** Cato Major, Lælius, Brutus, sive de Senectute, de Amicitia, de Claris Oratoribus Dialogi. With Notes by W. BROWNRIGG SMITH, M.A., F.R.G.S. 2s.

16. **Livy:** History of Rome. Notes by H. YOUNG and W. B. SMITH, M.A. Part 1. Books i., ii., 1s. 6d.

16*. ———— Part 2. Books iii., iv., v., 1s. 6d.
17. ———— Part 3. Books xxi., xxii., 1s. 6d.

19. **Latin Verse Selections,** from Catullus, Tibullus, Propertius, and Ovid. Notes by W. B. DONNE, M.A., Trinity College, Cambridge. 2s.

20. **Latin Prose Selections,** from Varro, Columella, Vitruvius, Seneca, Quintilian, Florus, Velleius Paterculus, Valerius Maximus Suetonius, Apuleius, &c. Notes by W. B. DONNE, M.A. 2s.

21. **Juvenalis** Satiræ. With Prolegomena and Notes by T. H. S. ESCOTT, B.A., Lecturer on Logic at King's College, London. 2s.

GREEK.

14. Greek Grammar, in accordance with the Principles and Philological Researches of the most eminent Scholars of our own day. By HANS CLAUDE HAMILTON. 1s. 6d.

15,17. Greek Lexicon. Containing all the Words in General Use, with their Significations, Inflections, and Doubtful Quantities. By HENRY R. HAMILTON. Vol. 1. Greek-English, 2s. 6d.; Vol. 2. English-Greek, 2s. Or the Two Vols. in One, 4s. 6d.: cloth boards, 5s.

14,15. Greek Lexicon (as above). Complete, with the GRAMMAR, in 17. One Vol., cloth boards, 6s.

GREEK CLASSICS. With Explanatory Notes in English.

1. Greek Delectus. Containing Extracts from Classical Authors, with Genealogical Vocabularies and Explanatory Notes, by H. YOUNG. New Edition, with an improved and enlarged Supplementary Vocabulary, by JOHN HUTCHISON, M.A., of the High School, Glasgow. 1s. 6d.

2, 3. Xenophon's Anabasis; or, The Retreat of the Ten Thousand. Notes and a Geographical Register, by H. YOUNG. Part 1. Books i. to iii., 1s. Part 2. Books iv. to vii., 1s.

4. Lucian's Select Dialogues. The Text carefully revised, with Grammatical and Explanatory Notes, by H. YOUNG. 1s. 6d.

5-12. Homer, The Works of. According to the Text of BAEUMLEIN. With Notes, Critical and Explanatory, drawn from the best and latest Authorities, with Preliminary Observations and Appendices, by T. H. L. LEARY, M.A., D.C.L.

THE ILIAD.	Part 1. Books i. to vi., 1s. 6d.		Part 3. Books xiii. to xviii., 1s. 6d.	
	Part 2. Books vii. to xii., 1s. 6d.		Part 4. Books xix. to xxiv., 1s. 6d.	
THE ODYSSEY:	Part 1. Books i. to vi., 1s. 6d		Part 3. Books xiii. to xviii., 1s. 6d.	
	Part 2. Books vii. to xii., 1s. 6d.		Part 4. Books xix. to xxiv., and Hymns, 2s.	

13. Plato's Dialogues: The Apology of Socrates, the Crito, and the Phædo. From the Text of C. F. HERMANN. Edited with Notes, Critical and Explanatory, by the Rev. JAMES DAVIES, M.A. 2s.

14-17. Herodotus, The History of, chiefly after the Text of GAISFORD. With Preliminary Observations and Appendices, and Notes, Critical and Explanatory, by T. H. L. LEARY, M.A., D.C.L.
Part 1. Books i., ii. (The Clio and Euterpe), 2s.
Part 2. Books iii., iv. (The Thalia and Melpomene), 2s.
Part 3. Books v.-vii. (The Terpsichore, Erato, and Polymnia), 2s.
Part 4. Books viii., ix. (The Urania and Calliope) and Index, 1s. 6d.

18. Sophocles: Œdipus Tyrannus. Notes by H. YOUNG. 1s.

20. Sophocles: Antigone. From the Text of DINDORF. Notes, Critical and Explanatory, by the Rev. JOHN MILNER, B.A. 2s.

23. Euripides: Hecuba and Medea. Chiefly from the Text of DINDORF. With Notes, Critical and Explanatory, by W. BROWNRIGG SMITH, M.A., F.R.G.S. 1s. 6d.

26. Euripides: Alcestis. Chiefly from the Text of DINDORF. With Notes, Critical and Explanatory, by JOHN MILNER, B.A. 1s. 6d.

30. Æschylus: Prometheus Vinctus: The Prometheus Bound. From the Text of DINDORF. Edited, with English Notes, Critical and Explanatory, by the Rev. JAMES DAVIES, M.A. 1s.

32. Æschylus: Septem Contra Thebes: The Seven against Thebes. From the Text of DINDORF. Edited, with English Notes, Critical and Explanatory, by the Rev. JAMES DAVIES, M.A. 1s.

40. Aristophanes: Acharnians. Chiefly from the Text of C. H. WEISE. With Notes, by C. S. T. TOWNSHEND, M.A. 1s. 6d.

41. Thucydides: History of the Peloponnesian War. Notes by H. YOUNG. Book 1. 1s. 6d.

42. Xenophon's Panegyric on Agesilaus. Notes and Introduction by LL. F. W. JEWITT. 1s. 6d.

43. Demosthenes. The Oration on the Crown and the Philippics. With English Notes. By Rev. T. H. L. LEARY, D.C.L., formerly Scholar of Brasenose College, Oxford. 1s. 6d.

CROSBY LOCKWOOD AND SON, 7, STATIONERS' HALL COURT, E.C.

7, STATIONERS' HALL COURT, LONDON, E.C.
March, 1891.

A

CATALOGUE OF BOOKS

INCLUDING NEW AND STANDARD WORKS IN

ENGINEERING: CIVIL, MECHANICAL, AND MARINE,
MINING AND METALLURGY,
ELECTRICITY AND ELECTRICAL ENGINEERING,
ARCHITECTURE AND BUILDING,
INDUSTRIAL AND DECORATIVE ARTS, SCIENCE, TRADE,
AGRICULTURE, GARDENING,
LAND AND ESTATE MANAGEMENT, LAW, &c.

PUBLISHED BY

CROSBY LOCKWOOD & SON.

MECHANICAL ENGINEERING, etc.

New Manual for Practical Engineers.

THE PRACTICAL ENGINEER'S HAND-BOOK. Comprising a Treatise on Modern Engines and Boilers: Marine, Locomotive and Stationary. And containing a large collection of Rules and Practical Data relating to recent Practice in Designing and Constructing all kinds of Engines, Boilers, and other Engineering work. The whole constituting a comprehensive Key to the Board of Trade and other Examinations for Certificates of Competency in Modern Mechanical Engineering. By WALTER S. HUTTON, Civil and Mechanical Engineer, Author of "The Works' Manager's Handbook for Engineers," &c. With upwards of 370 Illustrations. Third Edition, Revised, with Additions. Medium 8vo, nearly 500 pp., price 18s. Strongly bound.

☞ *This work is designed as a companion to the Author's "WORKS' MANAGER'S HAND-BOOK." It possesses many new and original features, and contains, like its predecessor, a quantity of matter not originally intended for publication, but collected by the author for his own use in the construction of a great variety of modern engineering work.*

The information is given in a condensed and concise form, and is illustrated by upwards of 370 Woodcuts; and comprises a quantity of tabulated matter of great value to all engaged in designing, constructing, or estimating for ENGINES, BOILERS and OTHER ENGINEERING WORK.

*** OPINIONS OF THE PRESS.

" We have kept it at hand for several weeks, referring to it as occasion arose, and we have not on a single occasion consulted its pages without finding the information of which we were in quest."
—*Athenæum.*

" A thoroughly good practical handbook, which no engineer can go through without learning something that will be of service to him."—*Marine Engineer.*

" An excellent book of reference for engineers, and a valuable text-book for students of engineering."—*Scotsman.*

" This valuable manual embodies the results and experience of the leading authorities on mechanical engineering."—*Building News.*

" The author has collected together a surprising quantity of rules and practical data, and has shown much judgment in the selections he has made. . . . There is no doubt that this book is one of the most useful of its kind published, and will be a very popular compendium."—*Engineer.*

" A mass of information, set down in simple language, and in such a form that it can be easily referred to at any time. The matter is uniformly good and well chosen, and is greatly elucidated by the illustrations. The book will find its way on to most engineers' shelves, where it will rank as one of the most useful books of reference."—*Practical Engineer.*

" Full of useful information, and should be found on the office shelf of all practical engineers.'
—*English Mechanic.*

B

Handbook for Works' Managers.

THE WORKS' MANAGER'S HANDBOOK OF MODERN RULES, TABLES, AND DATA. For Engineers, Millwrights, and Boiler Makers; Tool Makers, Machinists, and Metal Workers; Iron and Brass Founders, &c. By W. S. HUTTON, Civil and Mechanical Engineer, Author of "The Practical Engineer's Handbook." Fourth Edition, carefully Revised, and partly Re-written. In One handsome Volume, medium 8vo, price 15s. strongly bound. *[Just published.*

☞ *The Author having compiled Rules and Data for his own use in a great variety of modern engineering work, and having found his notes extremely useful, decided to publish them—revised to date—believing that a practical work, suited to the* DAILY REQUIREMENTS OF MODERN ENGINEERS, *would be favourably received.*

In the Third Edition, the following among other additions have been made, viz.: Rules for the Proportions of Riveted Joints in Soft Steel Plates, the Results of Experiments by PROFESSOR KENNEDY *for the Institution of Mechanical Engineers—Rules for the Proportions of Turbines—Rules for the Strength of Hollow Shafts of Whitworth's Compressed Steel, &c.*

** OPINIONS OF THE PRESS.

" The author treats every subject from the point of view of one who has collected workshop notes for application in workshop practice, rather than from the theoretical or literary aspect. The volume contains a great deal of that kind of information which is gained only by practical experience, and is seldom written in books."—*Engineer.*

"The volume is an exceedingly useful one, brimful with engineers' notes, memoranda, and rules, and well worthy of being on every mechanical engineer's bookshelf."—*Mechanical World.*

"The information is precisely that likely to be required in practice. . . . The work forms a desirable addition to the library not only of the works manager, but of anyone connected with general engineering."—*Mining Journal.*

"A formidable mass of facts and figures, readily accessible through an elaborate index . . . . . Such a volume will be found absolutely necessary as a book of reference in all sorts of 'works' connected with the metal trades."—*Ryland's Iron Trades Circular.*

" Brimful of useful information, stated in a concise form, Mr. Hutton's books have met a pressing want among engineers. The book must prove extremely useful to every practical man possessing a copy."—*Practical Engineer.*

Practical Treatise on Modern Steam-Boilers.

STEAM-BOILER CONSTRUCTION. A Practical Handbook for Engineers, Boiler-Makers, and Steam Users. Containing a large Collection of Rules and Data relating to the Design, Construction, and Working of Modern Stationary, Locomotive, and Marine Steam-Boilers. By WALTER S. HUTTON, Civil and Mechanical Engineer, Author of "The Works' Manager's Handbook," "The Practical Engineer's Handbook," &c. With upwards of 300 Illustrations.

☞ *This work is written in the same style as Mr. Hutton's other practical Handbooks, which it is intended to supplement. It is in active preparation and will, it is expected, be ready in April.*

"The Modernised Templeton."

THE PRACTICAL MECHANIC'S WORKSHOP COMPANION. Comprising a great variety of the most useful Rules and Formulæ in Mechanical Science, with numerous Tables of Practical Data and Calculated Results for Facilitating Mechanical Operations. By WILLIAM TEMPLETON, Author of "The Engineer's Practical Assistant," &c. &c. Sixteenth Edition, Revised, Modernised, and considerably Enlarged by WALTER S. HUTTON, C.E., Author of "The Works' Manager's Handbook," "The Practical Engineer's Handbook," &c. Fcap. 8vo, nearly 500 pp., with Eight Plates and upwards of 250 Illustrative Diagrams, 6s., strongly bound for workshop or pocket wear and tear. *[Just published.*

** OPINIONS OF THE PRESS.

" In its modernised form Hutton's ' Templeton' should have a wide sale, for it contains much valuable information which the mechanic will often find of use, and not a few tables and notes which he might look for in vain in other works. This modernised edition will be appreciated by all who have learned to value the original editions of ' Templeton.'"—*English Mechanic.*

" It has met with great success in the engineering workshop, as we can testify ; and there are a great many men who, in a great measure, owe their rise in life to this little book."—*Building News.*

" This familiar text-book—well known to all mechanics and engineers—is of essential service to the every-day requirements of engineers, millwrights, and the various trades connected with engineering and building. The new modernised edition is worth its weight in gold."—*Building News.* (Second Notice.)

" This well-known and largely used book contains information, brought up to date, of the sort so useful to the foreman and draughtsman. So much fresh information has been introduced as to constitute it practically a new book. It will be largely used in the office and workshop."—*Mechanical World.*

" The publishers wisely entrusted the task of revision of this popular, valuable and useful book to Mr. Hutton, than whom a more competent man they could not have found."—*Iron.*

Stone-working Machinery.

STONE-WORKING MACHINERY, and the Rapid and Economical Conversion of Stone. With Hints on the Arrangement and Management of Stone Works. By M. POWIS BALE, M.I.M.E. With Illusts. Crown 8vo, 9s.
" The book should be in the hands of every mason or student of stone-work."—
Colliery Guardian.
"It is in every sense of the word a standard work upon a subject which the author is fully competent to deal exhaustively with."—*Builder's Weekly Reporter.*
"A capital handbook for all who manipulate stone for building or ornamental purposes."—*Machinery Market.*

Pump Construction and Management.

PUMPS AND PUMPING : A Handbook for Pump Users. Being Notes on Selection, Construction and Management. By M. POWIS BALE, M.I.M.E., Author of " Woodworking Machinery," " Saw Mills," &c. Crown 8vo, 2s. 6d. cloth. [*Just published.*
" The matter is set forth as concisely as possible. In fact, condensation rather than diffuseness has been the author's aim throughout; yet he does not seem to have omitted anything likely to be of use."—*Journal of Gas Lighting.*
" Thoroughly practical and simply and clearly written."—*Glasgow Herald.*

Turning.

LATHE-WORK : A Practical Treatise on the Tools, Appliances, and Processes employed in the Art of Turning. By PAUL N. HASLUCK. Fourth Edition, Revised and Enlarged. Cr. 8vo, 5s. cloth. [*Just published.*
" Written by a man who knows, not only how work ought to be done, but who also knows how to do it, and how to convey his knowledge to others. To all turners this book would be valuable."—*Engineering.*
" We can safely recommend the work to young engineers. To the amateur it will simply be invaluable. To the student it will convey a great deal of useful information."—*Engineer.*
"A compact, succinct, and handy guide to lathe-work did not exist in our language until Mr. Hasluck, by the publication of this treatise, gave the turner a true *vade-mecum.*"—*House Decorator*

Screw-Cutting.

SCREW THREADS : And Methods of Producing Them. With Numerous Tables, and complete directions for using Screw-Cutting Lathes. By PAUL N. HASLUCK, Author of " Lathe-Work," &c. With Fifty Illustrations. Third Edition, Revised and Enlarged. Waistcoat-pocket size, 1s. 6d. cloth. [*Just published.*
" Full of useful information, hints and practical criticism. Taps, dies and screwing-tools generally are illustrated and their action described."—*Mechanical World.*
" It is a complete compendium of all the details of the screw cutting lathe ; in fact a *multum-in-parvo* on all the subjects it treats upon."—*Carpenter and Builder.*

Smith's Tables for Mechanics, etc.

TABLES, MEMORANDA, AND CALCULATED RESULTS, FOR MECHANICS, ENGINEERS, ARCHITECTS, BUILDERS, etc. Selected and Arranged by FRANCIS SMITH. Fifth Edition, thoroughly Revised and Enlarged, with a New Section of ELECTRICAL TABLES, FORMULÆ, and MEMORANDA. Waistcoat-pocket size, 1s. 6d. limp leather. [*Just published.*
" It would, perhaps, be as difficult to make a small pocket-book selection of notes and formulæ to suit ALL engineers as it would be to make a universal medicine ; but Mr. Smith's waistcoat-pocket collection may be looked upon as a successful attempt."—*Engineer.*
" The best example we have ever seen of 250 pages of useful matter packed into the dimensions of a card-case."—*Building News.* "A veritable pocket treasury of knowledge."—*Iron.*

Engineer's and Machinist's Assistant.

THE ENGINEER'S, MILLWRIGHT'S, and MACHINIST'S PRACTICAL ASSISTANT. A collection of Useful Tables, Rules and Data. By WILLIAM TEMPLETON. 7th Edition, with Additions. 18mo, 2s. 6d. cloth.
" Occupies a foremost place among books of this kind. A more suitable present to an apprentice to any of the mechanical trades could not possibly be made."—*Building News.*
" A deservedly popular work, it should be in the 'drawer' of every mechanic."—*English Mechanic.*

Iron and Steel.

" IRON AND STEEL" : A Work for the Forge, Foundry, Factory, and Office. Containing ready, useful, and trustworthy Information for Iron-masters and their Stock-takers; Managers of Bar, Rail, Plate, and Sheet Rolling Mills; Iron and Metal Founders; Iron Ship and Bridge Builders; Mechanical, Mining, and Consulting Engineers; Architects, Contractors, Builders, and Professional Draughtsmen. By CHARLES HOARE, Author of " The Slide Rule," &c. Eighth Edition, Revised throughout and considerably Enlarged. 32mo, 6s. leather.
" For comprehensiveness the book has not its equal."—*Iron.*
" One of the best of the pocket books."—*English Mechanic.*
" We cordially recommend this book to those engaged in considering the details of all kinds Iron and steel works."—*Naval Science.*

Engineering Construction.

PATTERN-MAKING : A Practical Treatise, embracing the Main
Types of Engineering Construction, and including Gearing, both Hand and
Machine made, Engine Work, Sheaves and Pulleys, Pipes and Columns,
Screws, Machine Parts, Pumps and Cocks, the Moulding of Patterns in
Loam and Greensand, &c., together with the methods of Estimating the
weight of Castings; to which is added an Appendix of Tables for Workshop
Reference. By a FOREMAN PATTERN MAKER. With upwards of Three
Hundred and Seventy Illustrations. Crown 8vo, 7s. 6d. cloth.

"A well-written technical guide, evidently written by a man who understands and has practised what he has written about. . . . We cordially recommend it to engineering students, yɔu g
journeymen, and others desirous of being initiated into the mysteries of pattern-making."—*Builder.*
"We can confidently recommend this comprehensive treatise.'—*Building News.*
"Likely to prove a welcome guide to many workmen, especially to draughtsmen who have
lacked a training in the shops, pupils pursuing their practical studies in our factories, and to em-
ployers and managers in engineering works."—*Hardware Trade Journal.*
"More than 370 illustrations help to explain the text, which is, however, always clear and explicit, thus rendering the work an excellent *vade mecum* for the apprentice who desires to become
master of his trade."—*English Mechanic.*

Dictionary of Mechanical Engineering Terms.

LOCKWOOD'S DICTIONARY OF TERMS USED IN THE
PRACTICE OF MECHANICAL ENGINEERING, embracing those current
in the Drawing Office, Pattern Shop, Foundry, Fitting, Turning, Smith's and
Boiler Shops, &c. &c. Comprising upwards of 6,000 Definitions. Edited by
A FOREMAN PATTERN-MAKER, Author of "Pattern Making." Crown 8vo,
7s. 6d. cloth.

"Just the sort of handy dictionary required by the various trades engaged in mechanical en-
gineering. The practical engineering pupil will find the book of great value in his studies, and
every foreman engineer and mechanic should have a copy."—*Building News.*
"After a careful examination of the book, and trying all manner of words, we think that the
engineer will here find all he is likely to require. It will be largely used."—*Practical Engineer.*
"One of the most useful books which can be presented to a mechanic or student."—*English
Mechanic.*
"Not merely a dictionary, but, to a certain extent, also a most valuable guide. It strikes us as
a happy idea to combine with a definition of the phrase useful information on the subject of which
it treats."—*Machinery Market.*
"No word having connection with any branch of constructive engineering seems to be
omitted. No more comprehensive work has been, so far, issued."—*Knowledge.*
"We strongly commend this useful and reliable adviser to our friends in the workshop, and to
students everywhere."—*Colliery Guardian.*

Steam Boilers.

A TREATISE ON STEAM BOILERS: Their Strength, Con-
struction, and Economical Working. By ROBERT WILSON, C.E. Fifth Edition.
12mo, 6s. cloth.

"The best treatise that has ever been published on steam boilers."—*Engineer.*
"The author shows himself perfect master of his subject, and we heartily recommend all employing steam power to possess themselves of the work."—*Ryland's Iron Trade Circular.*

Boiler Chimneys.

BOILER AND FACTORY CHIMNEYS; Their Draught-Power
and Stability. With a Chapter on Lightning Conductors. By ROBERT
WILSON, A.I.C.E., Author of "A Treatise on Steam Boilers," &c. Second
Edition. Crown 8vo, 3s. 6d. cloth.

"Full of useful information, definite in statement, and thoroughly practical in treatment."—
The Local Government Chronicle.
"A valuable contribution to the literature of scientific building."—*The Builder.*

Boiler Making.

THE BOILER-MAKER'S READY RECKONER & ASSIST-
ANT. With Examples of Practical Geometry and Templating, for the Use
of Platers, Smiths and Riveters. By JOHN COURTNEY. Edited by D. K. CLARK,
M.I.C.E. Third Edition, 480 pp., with 140 Illusts. Fcap. 8vo, 7s. half-bound.

"A most useful work. . . . No workman or apprentice should be without this book."—
Iron Trade Circular.
"Boiler-makers will readily recognise the value of this volume. . . . The tables are clearly
printed, and so arranged that they can be referred to with the greatest facility, so that it cannot be
doubted that they will be generally appreciated and much used."—*Mining Journal.*

Warming.

HEATING BY HOT WATER; with Information and Sug-
gestions on the best Methods of Heating Public, Private and Horticultural
Buildings. By WALTER JONES. With upwards of 50 Illustrations, crown
8vo, 2s. cloth.

"We confidently recommend all interested in heating by hot water to secure a copy of this
valuable little treatise."—*The Plumber and Decorator.*

Steam Engine.

TEXT-BOOK ON THE STEAM ENGINE. With a Supplement on Gas Engines, and PART II. ON HEAT ENGINES. By T. M. GOODEVE, M.A., Barrister-at-Law, Professor of Mechanics at the Normal School of Science and the Royal School of Mines; Author of "The Principles of Mechanics," "The Elements of Mechanism," &c. Eleventh Edition, Enlarged. With numerous Illustrations. Crown 8vo, 6s. cloth. [*Just published.*

"Professor Goodeve has given us a treatise on the steam engine which will bear comparison with anything written by Huxley or Maxwell, and we can award it no higher praise."—*Engineer.*

" Mr. Goodeve's text-book is a work of which every young engineer should possess himself."—*Mining Journal.*

"Essentially practical in its aim. The manner of exposition leaves nothing to be desired."—*Scotsman.*

Gas Engines.

ON GAS-ENGINES. Being a Reprint, with some Additions, of the Supplement to the *Text-book on the Steam Engine*, by T. M. GOODEVE, M.A. Crown 8vo, 2s. 6d. cloth.

" Like all Mr. Goodeve's writings, the present is no exception in point of general excellence. It is a valuable little volume."—*Mechanical World.*

Steam.

THE SAFE USE OF STEAM. Containing Rules for Unprofessional Steam-users. By an ENGINEER. Sixth Edition. Sewed, 6d.

"If steam-users would but learn this little book by heart boiler explosions would become sensations by their rarity."—*English Mechanic.*

Office Book for Mechanical Engineers.

THE MECHANICAL ENGINEER'S REFERENCE BOOK, for Machine and Boiler Construction. In Two Parts. Part I. GENERAL ENGINEERING DATA. Part II. BOILER CONSTRUCTION. With 48 Plates and numerous Illustrations. By NELSON FOLEY, M.I.N.A. Folio, half-bound. Price £5 5s. [*Nearly ready.*

Coal and Speed Tables.

A POCKET BOOK OF COAL AND SPEED TABLES, for *Engineers and Steam-users.* By NELSON FOLEY, Author of "Boiler Construction." Pocket-size, 3s. 6d. cloth ; 4s. leather.

"These tables are designed to meet the requirements of every-day use ; they are of sufficient scope for most practical purposes, and may be commended to engineers and users of steam.."—*Iron.*

"This pocket-book well merits the attention of the practical engineer. Mr. Foley has compiled a very useful set of tables, the information contained in which is frequently required by engineers, coal consumers and users of steam."—*Iron and Coal Trades Review.*

Fire Engineering.

FIRES, FIRE-ENGINES, AND FIRE-BRIGADES. With a History of Fire-Engines, their Construction, Use, and Management ; Remarks on Fire-Proof Buildings, and the Preservation of Life from Fire ; Statistics of the Fire Appliances in English Towns ; Foreign Fire Systems ; Hints on Fire Brigades, &c. &c. By CHARLES F. T. YOUNG, C.E. With numerous Illustrations, 544 pp., demy 8vo, £1 4s. cloth.

" To such of our readers as are interested in the subject of fires and fire apparatus, we can most heartily commend this book. It is really the only English work we now have upon the subject."—*Engineering.*

"It displays much evidence of careful research ; and Mr. Young has put his facts neatly together. It is evident enough that his acquaintance with the practical details of the construction of steam fire engines, old and new, and the conditions with which it is necessary they should comply, is accurate and full."—*Engineer.*

Estimating for Engineering Work, &c.

ENGINEERING ESTIMATES, COSTS AND ACCOUNTS: A Guide to Commercial Engineering. With numerous Examples of Estimates and Costs of Millwright Work, Miscellaneous Productions, Steam Engines and Steam Boilers ; and a Section on the Preparation of Costs Accounts. By A GENERAL MANAGER. Demy 8vo, 12s. cloth. [*Just published.*

"This is an excellent and very useful book, covering subject m tter in constant requisition in every factory and workshop. . . . The book is invaluable, not only to the young engineer, but also to the estimate department of every works."—*Builder.*

"This book bears on every page evidence that it has been prepared by an engineer accustomed to the work, and is no mere compilation, but contains a mass of valuable information of a kind useful even to experienced engineers."—*Practical Engineer.*

"We accord the work unqualified praise. The information is given in a plain, straightforward manner, and bears throughout evidence of the intimate practical acquaintance of the author with every phrase of commercial engineering."—*Mechanical World.*

THE POPULAR WORKS OF MICHAEL REYNOLDS
("THE ENGINE DRIVER'S FRIEND").

Locomotive-Engine Driving.

LOCOMOTIVE-ENGINE DRIVING : *A Practical Manual for Engineers in charge of Locomotive Engines.* By MICHAEL REYNOLDS, Member of the Society of Engineers, formerly Locomotive Inspector L. B. and S. C. R. Eighth Edition. Including a KEY TO THE LOCOMOTIVE ENGINE. With Illustrations and Portrait of Author. Crown 8vo, 4s. 6d. cloth.

"Mr. Reynolds has supplied a want, and has supplied it well. We can confidently recommend the book, not only to the practical driver, but to everyone who takes an interest in the performance of locomotive engines."—*The Engineer.*

"Mr. Reynolds has opened a new chapter in the literature of the day. This admirable practical treatise, of the practical utility of which we have to speak in terms of warm commendation."—*Athenæum.*

"Evidently the work of one who knows his subject thoroughly."—*Railway Service Gazette.*

"Were the cautions and rules given in the book to become part of the every-day working of ur engine-drivers, we might have fewer distressing accidents to deplore."—*Scotsman.*

Stationary Engine Driving.

STATIONARY ENGINE DRIVING : *A Practical Manual for Engineers in charge of Stationary Engines.* By MICHAEL REYNOLDS. Fourth Edition, Enlarged. With Plates and Woodcuts. Crown 8vo, 4s. 6d. cloth.

"The author is thoroughly acquainted with his subjects, and his advice on the various points treated is clear and practical. . . . He has produced a manual which is an exceedingly useful one for the class for whom it is specially intended."—*Engineering.*

"Our author leaves no stone unturned. He is determined that his readers shall not only know something about the stationary engine, but all about it."—*Engineer.*

"An engineman who has mastered the contents of Mr. Reynolds's book will require but little actual experience with boilers and engines before he can be trusted to look after them."—*EnglishMechanic.*

The Engineer, Fireman, and Engine-Boy.

THE MODEL LOCOMOTIVE ENGINEER, FIREMAN, *and* ENGINE-BOY. Comprising a Historical Notice of the Pioneer Locomotive Engines and their Inventors. By MICHAEL REYNOLDS. With numerous Illustrations and a fine Portrait of George Stephenson. Crown 8vo, 4s. 6d. cloth.

"From the technical knowledge of the author it will appeal to the railway man of to-day more forcibly than anything written by Dr. Smiles. . . . The volume contains information of a technical kind, and facts that every driver should be familiar with."—*English Mechanic.*

"We should be glad to see this book in the possession of everyone in the kingdom who has ever laid, or is to lay, hands on a locomotive engine."—*Iron.*

Continuous Railway Brakes.

CONTINUOUS RAILWAY BRAKES : *A Practical Treatise on the several Systems in Use in the United Kingdom ; their Construction and Performance.* With copious Illustrations and numerous Tables. By MICHAEL REYNOLDS. Large crown 8vo, 9s. cloth.

"A popular explanation of the different brakes. It will be of great assistance in forming public opinion, and will be studied with benefit by those who take an interest in the brake."—*English Mechanic.*

"Written with sufficient technical detail to enable the principle and relative connection of the various parts of each particular brake to be readily grasped."—*Mechanical World.*

Engine-Driving Life.

ENGINE-DRIVING LIFE : *Stirring Adventures and Incidents in the Lives of Locomotive-Engine Drivers.* By MICHAEL REYNOLDS. Second Edition, with Additional Chapters. Crown 8vo, 2s. cloth.

"From first to last perfectly fascinating. Wilkie Collins's most thrilling conceptions are thrown into the shade by true incidents, endless in their variety, related in every page."—*North British Mail.*

"Anyone who wishes to get a real insight into railway life cannot do better than read 'Engine-Driving Life' for himself ; and if he once take it up he will find that the author's enthusiasm and real love of the engine-driving profession will carry him on till he has read every page."—*Saturday Review.*

Pocket Companion for Enginemen.

THE ENGINEMAN'S POCKET COMPANION AND PRAC-TICAL EDUCATOR FOR ENGINEMEN, BOILER ATTENDANTS, AND MECHANICS. By MICHAEL REYNOLDS. With Forty-five Illustrations and numerous Diagrams. Second Edition, Revised. Royal 18mo, 3s. 6d., strongly bound for pocket wear.

"This admirable work is well suited to accomplish its object, being the honest workmanship of a competent engineer."—*Glasgow Herald.*

"A most meritorious work, giving in a succinct and practical form all the information an engine-minder desirous of mastering the scientific principles of his daily calling would require."—*Miller.*

"A boon to those who are striving to become efficient mechanics."—*Daily Chronicle.*

French-English Glossary for Engineers, etc.

A POCKET GLOSSARY of TECHNICAL TERMS: ENGLISH-FRENCH, FRENCH-ENGLISH; with Tables suitable for the Architectural, Engineering, Manufacturing and Nautical Professions. By JOHN JAMES FLETCHER, Engineer and Surveyor. 200 pp. Waistcoat-pocket size, 1s. 6d., limp leather.

"It ought certainly to be in the waistcoat-pocket of every professional man."—*Iron.*

"It is a very great advantage for readers and correspondents in France and England to have so large a number of the words relating to engineering and manufacturers collected in a liliputian volume. The little book will be useful both to students and travellers."—*Architect.*

"The glossary of terms is very complete, and many of the tables are new and well arranged. We cordially commend the book."—*Mechanical World.*

Portable Engines.

THE PORTABLE ENGINE; ITS CONSTRUCTION AND MANAGEMENT. A Practical Manual for Owners and Users of Steam Engines generally. By WILLIAM DYSON WANSBROUGH. With 90 Illustrations. Crown 8vo, 3s. 6d. cloth.

"This is a work of value to those who use steam machinery. . . . Should be read by every-one who has a steam engine, on a farm or elsewhere."—*Mark Lane Express.*

"We cordially commend this work to buyers and owners of steam engines, and to those who have to do with their construction or use."—*Timber Trades Journal.*

"Such a general knowledge of the steam engine as Mr. Wansbrough furnishes to the reader should be acquired by all intelligent owners and others who use the steam engine."—*Building News.*

"An excellent text-book of this useful form of engine, which describes with all necessary minuteness the details of the various devices. . . 'The Hints to Purchasers' contain a good deal of commonsense and practical wisdom."—*English Mechanic.*

CIVIL ENGINEERING, SURVEYING, etc.

MR. HUMBER'S IMPORTANT ENGINEERING BOOKS.

The Water Supply of Cities and Towns.

A COMPREHENSIVE TREATISE on the WATER-SUPPLY OF CITIES AND TOWNS. By WILLIAM HUMBER, A-M.Inst.C.E., and M. Inst. M.E., Author of "Cast and Wrought Iron Bridge Construction," &c. &c. Illustrated with 50 Double Plates, 1 Single Plate, Coloured Frontispiece, and upwards of 250 Woodcuts, and containing 400 pages of Text. Imp. 4to, £6 6s. elegantly and substantially half-bound in morocco.

List of Contents.

I. Historical Sketch of some of the means that have been adopted for the Supply of Water to Cities and Towns.—II. Water and the Foreign Matter usually associated with it.—III. Rainfall and Evaporation.—IV. Springs and the water-bearing formations of various districts.—V. Measurement and Estimation of the flow of Water—VI. On the Selection of the Source of Supply.—VII. Wells.—VIII. Reservoirs.—IX. The Purification of Water.—X. Pumps.—XI. Pumping Machinery.—XII. | Conduits.—XIII. Distribution of Water.—XIV. Meters, Service Pipes, and House Fittings.—XV. The Law and Economy of Water Works.—XVI. Constant and Intermittent Supply.—XVII. Description of Plates.—Appendices, giving Tables of Rates of Supply, Velocities, &c. &c., together with Specifications of several Works illustrated, among which will be found: Aberdeen, Bideford, Canterbury, Dundee, Halifax, Lambeth, Rotherham, Dublin, and others.

"The most systematic and valuable work upon water supply hitherto produced in English, or in any other language. . . . Mr. Humber's work is characterised almost throughout by an exhaustiveness much more distinctive of French and German than of English technical treatises."—*Engineer.*

"We can congratulate Mr. Humber on having been able to give so large an amount of information on a subject so important as the water supply of cities and towns. The plates, fifty in number, are mostly drawings of executed works, and alone would have commanded the attention of every engineer whose practice may lie in this branch of the profession."—*Builder.*

Cast and Wrought Iron Bridge Construction.

A COMPLETE AND PRACTICAL TREATISE ON CAST AND WROUGHT IRON BRIDGE CONSTRUCTION, including Iron Foundations. In Three Parts—Theoretical, Practical, and Descriptive. By WILLIAM HUMBER, A.M.Inst.C.E., and M.Inst.M.E. Third Edition, Revised and much improved, with 115 Double Plates (20 of which now first appear in this edition), and numerous Additions to the Text. In Two Vols., imp. 4to, £6 16s. 6d. half-bound in morocco.

"A very valuable contribution to the standard literature of civil engineering. In addition to elevations, plans and sections, large scale details are given which very much enhance the instructive worth of those illustrations."—*Civil Engineer and Architect's Journal.*

"Mr. Humber's stately volumes, lately issued—in which the most important bridges erected during the last five years, under the direction of the late Mr. Brunel, Sir W. Cubitt, Mr. Hawkshaw, Mr. Page, Mr. Fowler, Mr. Hemans, and others among our most eminent engineers, are drawn and specified in great detail."—*Engineer*

MR. HUMBER'S GREAT WORK ON MODERN ENGINEERING.

Complete in Four Volumes, imperial 4to, price £12 12s., half-morocco. Each
Volume sold separately as follows:—

A RECORD OF THE PROGRESS OF MODERN ENGINEER-
ING. FIRST SERIES. Comprising Civil, Mechanical, Marine, Hydraulic,
Railway, Bridge, and other Engineering Works, &c. By WILLIAM HUMBER,
A-M.Inst.C.E., &c. Imp. 4to, with 36 Double Plates, drawn to a large scale,
Photographic Portrait of John Hawkshaw, C.E., F.R.S., &c., and copious
descriptive Letterpress, Specifications, &c., £3 3s. half-morocco.

List of the Plates and Diagrams.

Victoria Station and Roof, L. B. & S. C. R. (8 plates); Southport Pier (2 plates); Victoria Station and Roof, L. C. & D. and G. W. R. (6 plates); Roof of Cremorne Music Hall; Bridge over G. N. Railway; Roof of Station, Dutch Rhenish Rail (2 plates); Bridge over the | Thames, West London Extension Railway (5 plates); Armour Plates: Suspension Bridge, Thames (4 plates); The Allen Engine; Suspension Bridge, Avon (3 plates); Underground Railway (3 plates).

"Handsomely lithographed and printed. It will find favour with many who desire to preserve in a permanent form copies of the plans and specifications prepared for the guidance of the contractors for many important engineering works."—*Engineer.*

HUMBER'S RECORD OF MODERN ENGINEERING. SECOND
SERIES. Imp. 4to, with 36 Double Plates, Photographic Portrait of Robert
Stephenson, C.E., M.P., F.R.S., &c., and copious descriptive Letterpress,
Specifications, &c., £3 3s. half-morocco.

List of the Plates and Diagrams.

Birkenhead Docks, Low Water Basin (15 plates); Charing Cross Station Roof, C. C. Railway (3 plates); Digswell Viaduct, Great Northern Railway; Robbery Wood Viaduct, Great Northern Railway; Iron Permanent Way; Clydach Viaduct, Merthyr, Tredegar, | and Abergavenny Railway; Ebbw Viaduct, Merthyr, Tredegar, and Abergavenny Railway; College Wood Viaduct, Cornwall Railway; Dublin Winter Palace Roof (3 plates); Bridge over the Thames, L. C. & D. Railway (6 plates); Albert Harbour, Greenock (4 plates).

" Mr. Humber has done the profession good and true service, by the fine collection of examples he has here brought before the profession and the public."—*Practical Mechanic's Journal.*

HUMBER'S RECORD OF MODERN ENGINEERING. THIRD
SERIES. Imp. 4to, with 40 Double Plates, Photographic Portrait of J. R.
M'Clean, late Pres. Inst. C.E., and copious descriptive Letterpress, Speci-
fications, &c., £3 3s. half-morocco.

List of the Plates and Diagrams.

MAIN DRAINAGE, METROPOLIS.—*North Side.*—Map showing Interception of Sewers; Middle Level Sewer (2 plates); Outfall Sewer, Bridge over River Lea (3 plates); Outfall Sewer, Bridge over Marsh Lane, North Woolwich Railway, and Bow and Barking Railway Junction; Outfall Sewer, Bridge over Bow and Barking Railway (3 plates); Outfall Sewer, Bridge over East London Waterworks' Feeder (2 plates); Outfall Sewer, Reservoir (2 plates); Outfall Sewer, Tumbling Bay and Outlet; Outfall Sewer, Penstocks. *South Side.*—Outfall Sewer, Bermondsey Branch (2 plates); Outfall | Sewer, Reservoir and Outlet (4 plates); Outfall Sewer, Filth Hoist; Sections of Sewers (North and South Sides). THAMES EMBANKMENT.—Section of River Wall; Steamboat Pier, Westminster (2 plates); Landing Stairs between Charing Cross and Waterloo Bridges; York Gate (2 plates); Overflow and Outlet at Savoy Street Sewer (3 plates); Steamboat Pier, Waterloo Bridge (3 plates); Junction of Sewers, Plans and Sections; Gullies, Plans and Sections; Rolling Stock; Granite and Iron Forts.

" The drawings have a constantly increasing value, and whoever desires to possess clear representations of the two great works carried out by our Metropolitan Board will obtain Mr. Humber's volume."—*Engineer.*

HUMBER'S RECORD OF MODERN ENGINEERING. FOURTH
SERIES. Imp. 4to, with 36 Double Plates, Photographic Portrait of John
Fowler, late Pres. Inst. C.E., and copious descriptive Letterpress, Speci-
fications, &c., £3 3s. half-morocco.

List of the Plates and Diagrams.

Abbey Mills Pumping Station, Main Drainage, Metropolis (4 plates); Barrow Docks (5 plates); Manquis Viaduct, Santiago and Valparaiso Railway (2 plates); Adam's Locomotive, St. Helen's Canal Railway (2 plates); Cannon Street Station Roof, Charing Cross Railway (3 plates); Road Bridge over the River Moka (2 plates); Telegraphic Apparatus for | Mesopotamia; Viaduct over the River Wye, Midland Railway (3 plates); St. Germans Viaduct, Cornwall Railway (2 plates); Wrought-Iron Cylinder for Diving Bell; Millwall Docks (6 plates); Milroy's Patent Excavator; Metropolitan District Railway (6 plates); Harbours, Ports, and Breakwaters (3 plates).

" We gladly welcome another year's issue of this valuable publication from the able pen of Mr. Humber. The accuracy and general excellence of this work are well known, while its usefulness in giving the measurements and details of some of the latest examples of engineering, as carried out by the most eminent men in the profession, cannot be too highly prized."—*Artizan.*

MR. HUMBER'S ENGINEERING BOOKS—*continued.*

Strains, Calculation of.
A HANDY BOOK FOR THE CALCULATION OF STRAINS
IN GIRDERS AND SIMILAR STRUCTURES, AND THEIR STRENGTH.
Consisting of Formulæ and Corresponding Diagrams, with numerous details
for Practical Application, &c. By WILLIAM HUMBER, A-M.Inst.C.E., &c.
Fourth Edition. Crown 8vo, nearly 100 Woodcuts and 3 Plates, 7s. 6d. cloth.

" The formulæ are neatly expressed, and the diagrams good."—*Athenæum.*
"We heartily commend this really *handy* book to our engineer and architect readers."—*English Mechanic.*

Barlow's Strength of Materials, enlarged by Humber
A TREATISE ON THE STRENGTH OF MATERIALS;
with Rules for Application in Architecture, the Construction of Suspension
Bridges, Railways, &c. By PETER BARLOW, F.R.S. A New Edition, revised
by his Sons, P. W. BARLOW, F.R.S., and W. H. BARLOW, F.R.S.; to which
are added, Experiments by HODGKINSON, FAIRBAIRN, and KIRKALDY; and
Formulæ for Calculating Girders, &c. Arranged and Edited by W. HUMBER,
A-M.Inst.C.E. Demy 8vo, 400 pp., with 19 large Plates and numerous Woodcuts, 18s. cloth.

" Valuable alike to the student, tyro, and the experienced practitioner, it will always rank in
future, as it has hitherto done, as the standard treatise on that particular subject."—*Engineer.*
" There is no greater authority than Barlow."—*Building News.*
" As a scientific work of the first class, it deserves a foremost place on the bookshelves of every
civil engineer and practical mechanic."—*English Mechanic.*

Trigonometrical Surveying.
AN OUTLINE OF THE METHOD OF CONDUCTING A
*TRIGONOMETRICAL SURVEY, for the Formation of Geographical and
Topographical Maps and Plans, Military Reconnaissance, Levelling, &c.,* with
Useful Problems, Formulæ, and Tables. By Lieut.-General FROME, R.E.
Fourth Edition, Revised and partly Re-written by Major General Sir CHARLES
WARREN, G.C.M.G., R.E. With 19 Plates and 115 Woodcuts, royal 8vo, 16s.
cloth.

" The simple fact that a fourth edition has been called for is the best testimony to its merits.
No words of praise from us can strengthen the position so well and so steadily maintained by this
work. Sir Charles Warren has revised the entire work, and made such additions as were necessary
to bring every portion of the contents up to the present date."—*Broad Arrow.*

Field Fortification.
A TREATISE ON FIELD FORTIFICATION, THE ATTACK
OF FORTRESSES, MILITARY MINING, AND RECONNOITRING. By
Colonel I. S. MACAULAY, late Professor of Fortification in the R.M.A., Wool-
wich. Sixth Edition, crown 8vo, cloth, with separate Atlas of 12 Plates, 12s.

Oblique Bridges.
A PRACTICAL AND THEORETICAL ESSAY ON OBLIQUE
BRIDGES. With 13 large Plates. By the late GEORGE WATSON BUCK,
M.I.C.E. Third Edition, revised by his Son, J. H. WATSON BUCK, M.I.C.E.;
and with the addition of Description to Diagrams for Facilitating the Construction of Oblique Bridges, by W. H. BARLOW, M.I.C.E. Royal 8vo, 12s.
cloth.

" The standard text-book for all engineers regarding skew arches is Mr. Buck's treatise, and it
would be impossible to consult a better."—*Engineer.*
"Mr. Buck's treatise is recognised as a standard text-book, and his treatment has divested the
subject of many of the intricacies supposed to belong to it. As a guide to the engineer and architect, on a confessedly difficult subject, Mr. Buck's work is unsurpassed."—*Building News.*

Water Storage, Conveyance and Utilisation.
WATER ENGINEERING : A Practical Treatise on the Measure-
ment, Storage, Conveyance and Utilisation of Water for the Supply of Towns,
for Mill Power, and for other Purposes. By CHARLES SLAGG, Water and
Drainage Engineer, A.M.Inst.C.E., Author of " Sanitary Work in the Smaller
Towns, and in Villages," &c. With numerous Illusts. Cr. 8vo, 7s. 6d. cloth.

" As a small practical treatise on the water supply of towns, and on some applications of
water-power, the work is in many respects excellent."—*Engineering.*
" The author has collated the results deduced from the experiments of the most eminent
authorities, and has presented them in a compact and practical form, accompanied by very clear
and detailed explanations. . . . The application of water as a motive power is treated very
carefully and exhaustively."—*Builder.*
" For anyone who desires to begin the study of hydraulics with a consideration of the practical
applications of the science there is no better guide."—*Architect.*

Statics, Graphic and Analytic.

GRAPHIC AND ANALYTIC STATICS, in their Practical Application to the Treatment of Stresses in Roofs, Solid Girders, Lattice, Bowstring and Suspension Bridges, Braced Iron Arches and Piers, and other Frameworks. By R. HUDSON GRAHAM, C.E. Containing Diagrams and Plates to Scale. With numerous Examples, many taken from existing Structures. Specially arranged for Class-work in Colleges and Universities. Second Edition, Revised and Enlarged. 8vo, 16s. cloth.

" Mr. Graham's book will find a place wherever graphic and analytic statics are used or studied."
—*Engineer.*

" The work is excellent from a practical point of view, and has evidently been prepared with much care. The directions for working are ample, and are illustrated by an abundance of well-selected examples. It is an excellent text-book for the practical draughtsman."—*Athenæum.*

Student's Text-Book on Surveying.

PRACTICAL SURVEYING : A Text-Book for Students preparing for Examination or for Survey-work in the Colonies. By GEORGE W. USILL, A.M.I.C.E., Author of "The Statistics of the Water Supply of Great Britain." With Four Lithographic Plates and upwards of 330 Illustrations. Second Edition, Revised. Crown 8vo, 7s. 6d. cloth. [*Just published.*

" The best forms of instruments are described as to their construction, uses and modes of employment, and there are innumerable hints on work and equipment such as the author, in his experience as surveyor, draughtsman and teacher, has found necessary, and which the student in his inexperience will find most serviceable."—*Engineer.*

" The latest treatise in the English language on surveying, and we have no hesitation in saying that the student will find it a better guide than any of its predecessors . . . . . Deserves to be recognised as the first book which should be put in the hands of a pupil of Civil Engineering, and every gentleman of education who sets out for the Colonies would find it well to have a copy."—*Architect.*

" A very useful, practical handbook on field practice. Clear, accurate and not too condensed."—*Journal of Education.*

Survey Practice.

AID TO SURVEY PRACTICE, for Reference in Surveying, Levelling, and Setting-out ; and in Route Surveys of Travellers by Land and Sea. With Tables, Illustrations, and Records. By LOWIS D'A. JACKSON, A.M.I.C.E., Author of "Hydraulic Manual," "Modern Metrology," &c. Second Edition, Enlarged. Large crown 8vo, 12s. 6d. cloth.

" Mr. Jackson has produced a valuable *vade-mecum* for the surveyor. We can recommend this book as containing an admirable supplement to the teaching of the accomplished surveyor."—*Athenæum.*

" As a text-book we should advise all surveyors to place it in their libraries, and study well the matured instructions afforded in its pages."—*Colliery Guardian.*

" The author brings to his work a fortunate union of theory and practical experience which, aided by a clear and lucid style of writing, renders the book a very useful one."—*Builder.*

Surveying, Land and Marine.

LAND AND MARINE SURVEYING, in Reference to the Preparation of Plans for Roads and Railways; Canals, Rivers, Towns' Water Supplies; Docks and Harbours. With Description and Use of Surveying Instruments. By W. D. HASKOLL, C.E., Author of "Bridge and Viaduct Construction," &c. Second Edition, Revised, with Additions. Large cr. 8vo, 9s. cl.

" This book must prove of great value to the student. We have no hesitation in recommending it, feeling assured that it will more than repay a careful study."—*Mechanical World.*

" A most useful and well arranged book for the aid of a student. We can strongly recommend it as a carefully-written and valuable text-book. It enjoys a well-deserved repute among surveyors."
—*Builder.*

" This volume cannot fail to prove of the utmost practical utility. It may be safely recommended to all students who aspire to become clean and expert surveyors."—*Mining Journal.*

Tunnelling.

PRACTICAL TUNNELLING. Explaining in detail the Setting-out of the works, Shaft-sinking and Heading-driving, Ranging the Lines and Levelling underground, Sub-Excavating, Timbering, and the Construction of the Brickwork of Tunnels, with the amount of Labour required for, and the Cost of, the various portions of the work. By FREDERICK W. SIMMS, F.G.S., M.Inst.C.E. Third Edition, Revised and Extended by D. KINNEAR CLARK, M.Inst.C.E. Imperial 8vo, with 21 Folding Plates and numerous Wood Engravings, 30s. cloth.

" The estimation in which Mr. Simms's book on tunnelling has been held for over thirty years cannot be more truly expressed than in the words of the late Prof. Rankine :—'The best source of information on the subject of tunnels is Mr. F. W. Simms's work on Practical Tunnelling.'"—*Architect.*

" It has been regarded from the first as a text book of the subject. . . . Mr. Clarke has added immensely to the value of the book."—*Engineer*

Levelling.

A TREATISE ON THE PRINCIPLES AND PRACTICE OF LEVELLING. Showing its Application to purposes of Railway and Civil Engineering, in the Construction of Roads; with Mr. TELFORD's Rules for the same. By FREDERICK W. SIMMS, F.G.S., M.Inst.C.E. Seventh Edition, with the addition of LAW's Practical Examples for Setting-out Railway Curves, and TRAUTWINE's Field Practice of Laying-out Circular Curves. With 7 Plates and numerous Woodcuts, 8vo, 8s. 6d., cloth. *₊* TRAUTWINE on Curves may be had separate, 5s.

" The text-book on levelling in most of our engineering schools and colleges."—*Engineer.*
" The publishers have rendered a substantial service to the profession, especially to the younger members, by bringing out the present edition of Mr. Simms's useful work."—*Engineering.*

Heat, Expansion by.

EXPANSION OF STRUCTURES BY HEAT. By JOHN KEILY, C.E., late of the Indian Public Works and Victorian Railway Departments. Crown 8vo, 3s. 6d. cloth.

SUMMARY OF CONTENTS.

Section I. FORMULAS AND DATA.	Section VI. MECHANICAL FORCE OF HEAT.
Section II. METAL BARS.	
Section III. SIMPLE FRAMES.	Section VII. WORK OF EXPANSION AND CONTRACTION.
Section IV. COMPLEX FRAMES AND PLATES.	
Section V. THERMAL CONDUCTIVITY.	Section VIII. SUSPENSION BRIDGES.
	Section IX. MASONRY STRUCTURES.

" The aim the author has set before him, viz., to show the effects of heat upon metallic and other structures, is a laudable one, for this is a branch of physics upon which the engineer or architect can find but little reliable and comprehensive data in books."—*Builder.*
" Whoever is concerned to know the effect of changes of temperature on such structures as suspension bridges and the like, could not do better than consult Mr. Keily's valuable and handy exposition of the geometrical principles involved in these changes."—*Scotsman.*

Practical Mathematics.

MATHEMATICS FOR PRACTICAL MEN: Being a Commonplace Book of Pure and Mixed Mathematics. Designed chiefly for the use of Civil Engineers, Architects and Surveyors. By OLINTHUS GREGORY, LL.D., F.R.A.S., Enlarged by HENRY LAW, C.E. 4th Edition, carefully Revised by J. R. YOUNG, formerly Professor of Mathematics, Belfast College. With 13 Plates, 8vo, £1 1s. cloth.

" The engineer or architect will here find ready to his hand rules for solving nearly every mathematical difficulty that may arise in his practice. The rules are in all cases explained by means of examples, in which every step of the process is clearly worked out."—*Builder.*
" One of the most serviceable books for practical mechanics. . . . It is an instructive book for the student, and a text-book for him who, having once mastered the subjects it treats of, needs occasionally to refresh his memory upon them."—*Building News.*

Hydraulic Tables.

HYDRAULIC TABLES, CO-EFFICIENTS, and FORMULÆ for finding the Discharge of Water from Orifices, Notches, Weirs, Pipes, and Rivers. With New Formulæ, Tables, and General Information on Rainfall, Catchment-Basins, Drainage, Sewerage, Water Supply for Towns and Mill Power. By JOHN NEVILLE, Civil Engineer, M.R.I.A. Third Ed., carefully Revised, with considerable Additions. Numerous Illusts. Cr. 8vo, 14s. cloth.

" Alike valuable to students and engineers in practice ; its study will prevent the annoyance of avoidable failures, and assist them to select the readiest means of successfully carrying out any given work connected with hydraulic engineering."—*Mining Journal.*
" It is, of all English books on the subject, the one nearest to completeness, . . . From the good arrangement of the matter, the clear explanations, and abundance of formulæ, the carefully calculated tables, and, above all, the thorough acquaintance with both theory and construction, which is displayed from first to last, the book will be found to be an acquisition."—*Architect.*

Hydraulics.

HYDRAULIC MANUAL. Consisting of Working Tables and Explanatory Text. Intended as a Guide in Hydraulic Calculations and Field Operations. By LOWIS D'A. JACKSON, Author of "Aid to Survey Practice," " Modern Metrology," &c. Fourth Edition, Enlarged. Large cr. 8vo, 16s. cl.

" The author has had a wide experience in hydraulic engineering and has been a careful observer of the facts which have come under his notice, and from the great mass of material at his command he has constructed a manual which may be accepted as a trustworthy guide to this branch of the engineer's profession. We can heartily recommend this volume to all who desire to be acquainted with the latest development of this important subject."—*Engineering.*
" The standard-work in this department of mechnnics."—*Scotsman.*
" The most useful feature of this work is its freedom from what is superannuated, and its thorough adoption of recent experiments; the text is, in fact, in great part a short account of the great modern experiments."—*Nature.*

Drainage.

ON THE DRAINAGE OF LANDS, TOWNS AND BUILD-INGS. By G. D. DEMPSEY, C.E., Author of "The Practical Railway Engineer," &c. Revised, with large Additions on RECENT PRACTICE IN DRAINAGE ENGINEERING, by D. KINNEAR CLARK, M.Inst.C.E. Author of "Tramways: Their Construction and Working," "A Manual of Rules, Tables, and Data for Mechanical Engineers," &c, &c. Crown 8vo, 7s. 6d. cloth.

[*Just published.*

" The new matter added to Mr. Dempsey's excellent work is characterised by the comprehensive grasp and accuracy of detail for which the name of Mr. D. K. Clark is a sufficient voucher."—*Athenæum.*

" As a work on recent practice in drainage engineering, the book is to be commended to all who are making that branch of engineering science their special study."—*Iron.*

" A comprehensive manual on drainage engineering, and a useful introduction to the student." *Building News.*

Tramways and their Working.

TRAMWAYS: THEIR CONSTRUCTION AND WORKING. Embracing a Comprehensive History of the System; with an exhaustive Analysis of the various Modes of Traction, including Horse-Power, Steam, Heated Water, and Compressed Air; a Description of the Varieties of Rolling Stock; and ample Details of Cost and Working Expenses: the Progress recently made in Tramway Construction, &c. &c. By D. KINNEAR CLARK, M.Inst.C.E. With over 200 Wood Engravings, and 13 Folding Plates. Two Vols., large crown 8vo, 30s. cloth.

" All interested in tramways must refer to it, as all railway engineers have turned to the author's work 'Railway Machinery.'"—*Engineer.*

" An exhaustive and practical work on tramways, in which the history of this kind of locomotion, and a description and cost of the various modes of laying tramways, are to be found."—*Building News.*

" The best form of rails, the best mode of construction, and the best mechanical appliances are so fairly indicated in the work under review, that any engineer about to construct a tramway will be enabled at once to obtain the practical information which will be of most service to him."—*Athenæum.*

Oblique Arches.

A PRACTICAL TREATISE ON THE CONSTRUCTION OF OBLIQUE ARCHES. By JOHN HART. Third Edition, with Plates. Imperial 8vo, 8s. cloth.

Curves, Tables for Setting-out.

TABLES OF TANGENTIAL ANGLES AND MULTIPLES *for Setting-out Curves from 5 to 200 Radius.* By ALEXANDER BEAZELEY, M.Inst.C.E. Third Edition. Printed on 48 Cards, and sold in a cloth box, waistcoat-pocket size, 3s. 6d.

" Each table is printed on a small card, which, being placed on the theodolite, leaves the hands free to manipulate the instrument—no small advantage as regards the rapidity of work."—*Engineer.*

" Very handy; a man may know that all his day's work must fall on two of these cards, which he puts into his own card-case, and leaves the rest behind."—*Athenæum.*

Earthwork.

EARTHWORK TABLES. Showing the Contents in Cubic Yards of Embankments, Cuttings, &c., of Heights or Depths up to an average of 80 feet. By JOSEPH BROADBENT, C.E., and FRANCIS CAMPIN, C.E. Crown 8vo, 5s. cloth.

" The way in which accuracy is attained, by a simple division of each cross section into three elements, two in which are constant and one variable, is ingenious."—*Athenæum.*

Tunnel Shafts.

THE CONSTRUCTION OF LARGE TUNNEL SHAFTS: *A Practical and Theoretical Essay.* By J. H. WATSON BUCK, M.Inst.C.E., Resident Engineer, London and North-Western Railway. Illustrated with Folding Plates, royal 8vo, 12s. cloth.

" Many of the methods given are of extreme practical value to the mason; and the observations on the form of arch, the rules for ordering the stone, and the construction of the templates will be found of considerable use. We commend the book to the engineering profession."—*Building News.*

" Will be regarded by civil engineers as of the utmost value, and calculated to save much time and obviate many mistakes."—*Colliery Guardian.*

Girders, Strength of.

GRAPHIC TABLE FOR FACILITATING THE COMPUTATION OF THE WEIGHTS OF WROUGHT IRON AND STEEL GIRDERS, etc., for Parliamentary and other Estimates. By J. H. WATSON BUCK, M.Inst.C.E. On a Sheet, 2s. 6d.

River Engineering.

RIVER BARS: The Causes of their Formation, and their Treatment by " Induced Tidal Scour;" with a Description of the Successful Reduction by this Method of the Bar at Dublin. By I. J. MANN, Assist. Eng. to the Dublin Port and Docks Board. Royal 8vo, 7s. 6d. cloth.

"We recommend all interested in harbour works—and, indeed, those concerned in the improvements of rivers generally—to read Mr. Mann's interesting work on the treatment of river bars."—Engineer.

Trusses.

TRUSSES OF WOOD AND IRON. Practical Applications of Science in Determining the Stresses, Breaking Weights, Safe Loads, Scantlings, and Details of Construction, with Complete Working Drawings. By WILLIAM GRIFFITHS, Surveyor, Assistant Master, Tranmere School of Science and Art. Oblong 8vo, 4s. 6d. cloth.

" This handy little book enters so minutely into every detail connected with the construction of roof trusses, that no student need be ignorant of these matters."—Practical Engineer.

Railway Working.

SAFE RAILWAY WORKING. A Treatise on Railway Accidents: Their Cause and Prevention; with a Description of Modern Appliances and Systems. By CLEMENT E. STRETTON, C.E., Vice-President and Consulting Engineer, Amalgamated Society of Railway Servants. With Illustrations and Coloured Plates. Second Edition, Enlarged. Crown 8vo, 3s. 6d. cloth. [Just published.

" A book for the engineer, the directors, the managers; and, in short, all who wish for information on railway matters will find a perfect encyclopædia in 'Safe Railway Working.'"—Railway Review.

" We commend the remarks on railway signalling to all railway managers, especially where a uniform code and practice is advocated."—Herepath's Railway Journal.

" The author may be congratulated on having collected, in a very convenient form, much valuable information on the principal questions affecting the safe working of railways."—Railway Engineer.

Field-Book for Engineers.

THE ENGINEER'S, MINING SURVEYOR'S, AND CONTRACTOR'S FIELD-BOOK. Consisting of a Series of Tables, with Rules, Explanations of Systems, and use of Theodolite for Traverse Surveying and Plotting the Work with minute accuracy by means of Straight Edge and Set Square only; Levelling with the Theodolite, Casting-out and Reducing Levels to Datum, and Plotting Sections in the ordinary manner; setting-out Curves with the Theodolite by Tangential Angles and Multiples, with Right and Left-hand Readings of the Instrument: Setting-out Curves without Theodolite, on the System of Tangential Angles by sets of Tangents and Offsets; and Earthwork Tables to 80 feet deep, calculated for every 6 inches in depth. By W. DAVIS HASKOLL, C.E. With numerous Woodcuts. Fourth Edition, Enlarged. Crown 8vo, 12s. cloth.

"The book is very handy; the separate tables of sines and tangents to every minute will make t useful for many other purposes, the genuine traverse tables existing all the same."—Athenæum.

"Every person engaged in engineering field operations will estimate the importance of such a work and the amount of valuable time which will be saved by reference to a set of reliable tables prepared with the accuracy and fulness of those given in this volume."—Railway News.

Earthwork, Measurement of.

A MANUAL ON EARTHWORK. By ALEX. J. S. GRAHAM, C.E. With numerous Diagrams. Second Edition. 18mo, 2s. 6d. cloth.

"A great amount of practical information, very admirably arranged, and available for rough estimates, as well as for the more exact calculations required in the engineer's and contractor's offices."—Artisan.

Strains in Ironwork.

THE STRAINS ON STRUCTURES OF IRONWORK; with Practical Remarks on Iron Construction. By F. W. SHEILDS, M.Inst.C.E. Second Edition, with 5 Plates. Royal 8vo, 5s. cloth.

"The student cannot find a better little book on this subject."—Engineer.

Cast Iron and other Metals, Strength of.

A PRACTICAL ESSAY ON THE STRENGTH OF CAST IRON AND OTHER METALS. By THOMAS TREDGOLD, C.E. Fifth Edition, including HODGKINSON'S Experimental Researches. 8vo, 12s. cloth.

ARCHITECTURE, BUILDING, etc.

Construction.
THE SCIENCE OF BUILDING : An Elementary Treatise on the Principles of Construction. By E. WYNDHAM TARN, M.A., Architect. Third Edition, Revised and Enlarged, with 59 Engravings. Fcap. 8vo, 4s. cloth. [Just published.
"A very valuable book, which we strongly recommend to all students."—Builder.
"No architectural student should be without this handbook of constructional knowledge."—Architect.

Villa Architecture.
A HANDY BOOK OF VILLA ARCHITECTURE : Being a Series of Designs for Villa Residences in various Styles. With Outline Specifications and Estimates. By C. WICKES, Architect, Author of "The Spires and Towers of England," &c. 61 Plates, 4to, £1 11s. 6d. half-morocco, gilt edges.
"The whole of the designs bear evidence of their being the work of an artistic architect, and they will prove very valuable and suggestive."—Building News.

Text-Book for Architects.
THE ARCHITECT'S GUIDE: Being a Text-Book of Useful Information for Architects, Engineers, Surveyors, Contractors, Clerks of Works, &c. &c. By FREDERICK ROGERS, Architect, Author of " Specifications for Practical Architecture," &c. Second Edition, Revised and Enlarged. With numerous Illustrations. Crown 8vo, 6s. cloth.
"As a text-book of useful information for architects, engineers, surveyors, &c., it would be hard to find a handier or more complete little volume."—Standard.
"A young architect could hardly have a better guide-book."—Timber Trades Journal.

Taylor and Cresy's Rome.
THE ARCHITECTURAL ANTIQUITIES OF ROME. By the late G. L. TAYLOR, Esq., F.R.I.B.A., and EDWARD CRESY, Esq. New Edition, thoroughly Revised by the Rev. ALEXANDER TAYLOR, M.A. (son of the late G. L. Taylor, Esq.), Fellow of Queen's College, Oxford, and Chaplain of Gray's Inn. Large folio, with 130 Plates, half-bound, £3 3s.
N.B.—This is the only book which gives on a large scale, and with the precision of architectural measurement, the principal Monuments of Ancient Rome in plan, elevation, and detail.
Taylor and Cresy's work has from its first publication been ranked among those professional books which cannot be bettered. . . . It would be difficult to find examples of drawings, even among those of the most painstaking students of Gothic, more thoroughly worked out than are the one hundred and thirty plates in this volume."—Architect.

Architectural Drawing.
PRACTICAL RULES ON DRAWING, for the Operative Builder and Young Student in Architecture. By GEORGE PYNE. With 14 Plates, 4to, 7s. 6d. boards.

Sir Wm. Chambers's Treatise on Civil Architecture.
THE DECORATIVE PART OF CIVIL ARCHITECTURE. By Sir WILLIAM CHAMBERS, F.R.S. With Portrait, Illustrations, Notes, and an Examination of Grecian Architecture, by JOSEPH GWILT, F.S.A. Revised and Edited by W. H. LEEDS, with a Memoir of the Author. 66 Plates, 4to, 21s. cloth.

House Building and Repairing.
THE HOUSE-OWNER'S ESTIMATOR ; or, What will it Cost to Build, Alter, or Repair? A Price Book adapted to the Use of Unprofessional People, as well as for the Architectural Surveyor and Builder. By JAMES D. SIMON, A.R.I.B.A. Edited and Revised by FRANCIS T. W. MILLER, A.R.I.B.A. With numerous Illustrations, Fourth Edition, Revised. Crown 8vo, 3s. 6d. cloth.
"In two years it will repay its cost a hundred times over."—Field.
"A very handy book."—English Mechanic.

Cottages and Villas.
COUNTRY AND SUBURBAN COTTAGES AND VILLAS: How to Plan and Build Them. Containing 33 Plates, with Introduction, General Explanations, and Description of each Plate. By JAMES W. BOGUE, Architect, Author of "Domestic Architecture," &c. 4to, 10s. 6d. cloth.
[Just published.

The New Builder's Price Book, 1891.

LOCKWOOD'S BUILDER'S PRICE BOOK FOR 1891. A Comprehensive Handbook of the Latest Prices and Data for Builders, Architects, Engineers and Contractors. Re-constructed, Re-written and Greatly Enlarged. By FRANCIS T. W. MILLER. 640 closely-printed pages, crown 8vo, 4s. cloth. [*Just published.*
"This book is a very useful one, and should find a place in every English office connected with the building and engineering professions."—*Industries.*
"This Price Book has been set up in new type. . . . Advantage has been taken of the transformation to add much additional information, and the volume is now an excellent book of reference."—*Architect.*
"In its new and revised form this Price Book is what a work of this kind should be—comprehensive, reliable, well arranged, legible and well bound."—*British Architect.*
"A work of established reputation."—*Athenæum.*
"This very useful handbook is well written, exceedingly clear in its explanations and great care has evidently been taken to ensure accuracy."—*Morning Advertiser*

Designing, Measuring, and Valuing.

THE STUDENT'S GUIDE to the PRACTICE of MEASUR-ING AND VALUING ARTIFICERS' WORKS. Containing Directions for taking Dimensions, Abstracting the same, and bringing the Quantities into Bill, with Tables of Constants for Valuation of Labour, and for the Calculation of Areas and Solidities. Originally edited by EDWARD DOBSON, Architect. With Additions on Mensuration and Construction, and a New Chapter on Dilapidations, Repairs, and Contracts, by E. WYNDHAM TARN, M.A. Sixth Edition, including a Complete Form of a Bill of Quantities. With 8 Plates and 63 Woodcuts. Crown 8vo, 7s. 6d. cloth.
"Well fulfils the promise of its title-page, and we can thoroughly recommend it to the class for whose use it has been compiled. Mr. Tarn's additions and revisions have much increased the usefulness of the work, and have especially augmented its value to students."—*Engineering.*
"This edition will be found the most complete treatise on the principles of measuring and valuing artificers' work that has yet been published."—*Building News.*

Pocket Estimator and Technical Guide.

THE POCKET TECHNICAL GUIDE, MEASURER AND ESTIMATOR FOR BUILDERS AND SURVEYORS. Containing Technical Directions for Measuring Work in all the Building Trades, Complete Specifications for Houses, Roads, and Drains, and an easy Method of Estimating the parts of a Building collectively. By A. C. BEATON, Author of "Quantities and Measurements," &c. Fifth Edition. With 53 Woodcuts, waistcoat-pocket size, 1s. 6d. gilt edges.
"No builder, architect, surveyor, or valuer should be without his 'Beaton.'"—*Building News.*
"Contains an extraordinary amount of information in daily requisition in measuring and estimating. Its presence in the pocket will save valuable time and trouble."—*Building World.*

Donaldson on Specifications.

THE HANDBOOK OF SPECIFICATIONS; or, Practical Guide to the Architect, Engineer, Surveyor, and Builder, in drawing up Specifications and Contracts for Works and Constructions. Illustrated by Precedents of Buildings actually executed by eminent Architects and Engineers. By Professor T. L. DONALDSON, P.R.I.B.A., &c. New Edition, in One large Vol., 8vo, with upwards of 1,000 pages of Text, and 33 Plates, £1 11s. 6d. cloth.
"In this work forty-four specifications of executed works are given, including the specifications for parts of the new Houses of Parliament, by Sir Charles Barry, and for the new Royal Exchange, by Mr. Tite, M.P. The latter, in particular, is a very complete and remarkable document. It embodies, to a great extent, as Mr. Donaldson mentions, 'the bill of quantities with the description of the works.' . . . It is valuable as a record, and more valuable still as a book of precedents. . . . Suffice it to say that Donaldson's 'Handbook of Specifications' must be bought by all architects."—*Builder.*

Bartholomew and Rogers' Specifications.

SPECIFICATIONS FOR PRACTICAL ARCHITECTURE. A Guide to the Architect, Engineer, Surveyor, and Builder. With an Essay on the Structure and Science of Modern Buildings. Upon the Basis of the Work by ALFRED BARTHOLOMEW, thoroughly Revised, Corrected, and greatly added to by FREDERICK ROGERS, Architect. Second Edition, Revised, with Additions. With numerous Illustrations, medium 8vo, 15s. cloth.
"The collection of specifications prepared by Mr. Rogers on the basis of Bartholomew's work is too well known to need any recommendation from us. It is one of the books with which every young architect must be equipped ; for time has shown that the specifications cannot be set aside through any defect in them."—*Architect.*

Building ; Civil and Ecclesiastical.

A BOOK ON BUILDING, Civil and Ecclesiastical, including Church Restoration ; with the Theory of Domes and the Great Pyramid, &c. By Sir EDMUND BECKETT, Bart., LL.D., F.R.A.S., Author of " Clocks and Watches, and Bells," &c. Second Edition, Enlarged. Fcap. 8vo, 5s. cloth.
" A book which is always amusing and nearly always instructive. The style throughout is in the highest degree condensed and epigrammatic."—*Times.*

Ventilation of Buildings.

VENTILATION. A Text Book to the Practice of the Art of Ventilating Buildings. With a Chapter upon Air Testing. By W. P. BUCHAN, R.P., Sanitary and Ventilating Engineer, Author of " Plumbing," &c. With 170 Illustrations. 12mo, 4s. cloth boards.   [*Just published.*

The Art of Plumbing.

PLUMBING. A Text Book to the Practice of the Art or Craft of the Plumber, with Supplementary Chapters on House Drainage, embodying the latest Improvements. By WILLIAM PATON BUCHAN, R.P., Sanitary Engineer and Practical Plumber. Fifth Edition, Enlarged to 370 pages, and 380 Illustrations. 12mo, 4s. cloth boards.
" A text book which may be safely put in the hands of every young plumber, and which will also be found useful by architects and medical professors."—*Builder.*
" A valuable text book, and the only treatise which can be regarded as a really reliable manual of the plumber's art."—*Building News.*

Geometry for the Architect, Engineer, etc.

PRACTICAL GEOMETRY, for the Architect, Engineer and Mechanic. Giving Rules for the Delineation and Application of various Geometrical Lines, Figures and Curves. By E. W. TARN, M.A., Architect, Author of "The Science of Building," &c. Second Edition. With 172 Illustrations, demy 8vo, 9s. cloth.
" No book with the same objects in view has ever been published in which the clearness of the rules laid down and the illustrative diagrams have been so satisfactory."—*Scotsman.*

The Science of Geometry.

THE GEOMETRY OF COMPASSES; or, Problems Resolved by the mere Description of Circles, and the use of Coloured Diagrams and Symbols. By OLIVER BYRNE. Coloured Plates. Crown 8vo, 3s. 6d. cloth.
" The treatise is a good one, and remarkable—like all Mr. Byrne's contributions to the science of geometry—for the lucid character of its teaching."—*Building News.*

DECORATIVE ARTS, etc.

Woods and Marbles (Imitation of).

SCHOOL OF PAINTING FOR THE IMITATION OF WOODS AND MARBLES, as Taught and Practised by A. R. VAN DER BURG and P. VAN DER BURG, Directors of the Rotterdam Painting Institution. Royal folio, 18¼ by 12¼ in., Illustrated with 24 full-size Coloured Plates; also 12 plain Plates, comprising 154 Figures. Second and Cheaper Edition. Price £1 11s. 6d.

List of Plates.

1. Various Tools required for Wood Painting —2, 3. Walnut: Preliminary Stages of Graining and Finished Specimen — 4. Tools used for Marble Painting and Method of Manipulation—5, 6. St. Remi Marble: Earlier Operations and Finished Specimen—7. Methods of Sketching different Grains, Knots, &c.—8, 9. Ash: Preliminary Stages and Finished Specimen — 10. Methods of Sketching Marble Grains—11, 12. Breche Marble: Preliminary Stages of Working and Finished Specimen—13. Maple: Methods of Producing the different Grains—14, 15. Bird's-eye Maple: Preliminary Stages and Finished Specimen—16. Methods of Sketching the different Species of White Marble—17, 18. White Marble: Preliminary Stages of Process and Finished Specimen—19. Mahogany: Specimens of various Grains and Methods of Manipulation —20, 21. Mahogany: Earlier Stages and Finished Specimen—22, 23, 24. Sienna Marble: Varieties of Grain, Preliminary Stages and Finished Specimen—25, 26, 27. Juniper Wood: Methods of producing Grain, &c.; Preliminary Stages and Finished Specimen—28, 29, 30. Vert de Mer Marble: Varieties of Grain and Methods of Working Unfinished and Finished Specimens—31, 32, 33. Oak: Varieties of Grain, Tools Employed, and Methods of Manipulation, Preliminary Stages and Finished Specimen—34, 35, 36. Waulsort Marble: Varieties of Grain, Unfinished and Finished Specimens.

** OPINIONS OF THE PRESS.

" Those who desire to attain skill in the art of painting woods and marbles will find advantage in consulting this book. . . . Some of the Working Men's Clubs should give their young men the opportunity to study it."—*Builder.*
" A comprehensive guide to the art. The explanations of the processes, the manipulation and management of the colours, and the beautifully executed plates will not be the least valuable to the student who aims at making his work a faithful transcript of nature,"—*Building News.*

House Decoration.

ELEMENTARY DECORATION. A Guide to the Simpler Forms of Everyday Art, as applied to the Interior and Exterior Decoration of Dwelling Houses, &c. By JAMES W. FACEY, Jun. With 68 Cuts. 12mo, 2s. cloth limp.

PRACTICAL HOUSE DECORATION : A Guide to the Art of Ornamental Painting, the Arrangement of Colours in Apartments, and the principles of Decorative Design. With some Remarks upon the Nature and Properties of Pigments. By JAMES WILLIAM FACEY, Author of " Elementary Decoration," &c. With numerous Illustrations. 12mo, 2s. 6d. cloth limp.

N.B.—The above Two Works together in One Vol., strongly half-bound, 5s.

Colour.

A GRAMMAR OF COLOURING. Applied to Decorative Painting and the Arts. By GEORGE FIELD. New Edition, Revised, Enlarged, and adapted to the use of the Ornamental Painter and Designer. By ELLIS A. DAVIDSON. With New Coloured Diagrams and Engravings. 12mo, 3s. 6d. cloth boards.

"The book is a most useful *resume* of the properties of pigments."—*Builder.*

House Painting, Graining, etc.

HOUSE PAINTING, GRAINING, MARBLING, AND SIGN WRITING, A Practical Manual of. By ELLIS A. DAVIDSON. Fifth Edition. With Coloured Plates and Wood Engravings. 12mo, 6s. cloth boards.

" A mass of information, of use to the amateur and of value to the practical man."—*English Mechanic.*

"Simply invaluable to the youngster entering upon this particular calling, and highly service-able to the man who is practising it."—*Furniture Gazette.*

Decorators, Receipts for.

THE DECORATOR'S ASSISTANT : A Modern Guide to Decorative Artists and Amateurs, Painters, Writers, Gilders, &c. Containing upwards of 600 Receipts, Rules and Instructions ; with a variety of Information for General Work connected with every Class of Interior and Exterior Decorations, &c. Fourth Edition, Revised. 152 pp., crown 8vo, 1s. in wrapper.

" Full of receipts of value to decorators, painters, gilders, &c. The book contains the gist of larger treatises on colour and technical processes. It would be difficult to meet with a work so full of varied information on the painter's art."—*Building News.*

" We recommend the work to all who, whether for pleasure or profit, require a guide to decoration."—*Plumber and Decorator.*

Moyr Smith on Interior Decoration.

ORNAMENTAL INTERIORS, ANCIENT AND MODERN. By J. MOYR SMITH. Super-royal 8vo, with 32 full-page Plates and numerous smaller Illustrations, handsomely bound in cloth, gilt top, price 18s.

" The book is well illustrated and handsomely got up, and contains some true criticism and a good many good examples of decorative treatment."—*The Builder.*

"This is the most elaborate and beautiful work on the artistic decoration of interiors that we have seen. . . . The scrolls, panels and other designs from the author's own pen are very beautiful and chaste ; but he takes care that the designs of other men shall figure even more than his own."—*Liverpool Albion.*

" To all who take an interest in elaborate domestic ornament this handsome volume will be welcome."—*Graphic.*

British and Foreign Marbles.

MARBLE DECORATION and the Terminology of British and Foreign Marbles. A Handbook for Students. By GEORGE H. BLAGROVE, Author of " Shoring and its Application," &c. With 28 Illustrations. Crown 8vo, 3s. 6d. cloth.

" This most useful and much wanted handbook should be in the hands of every architect and builder."—*Building World.*

" It is an excellent manual for students, and interesting to artistic readers generally."—*Saturday Review.*

" A carefully and usefully written treatise ; the work is essentially practical."—*Scotsman.*

Marble Working, etc.

MARBLE AND MARBLE WORKERS : A Handbook for Architects, Artists, Masons and Students. By ARTHUR LEE, Author of " A Visit to Carrara," " The Working of Marble," &c. Small crown 8vo, 2s. cloth.

" A really valuable addition to the technical literature of architects and masons."—*Building News.*

C

DELAMOTTE'S WORKS ON ILLUMINATION AND ALPHABETS.

A PRIMER OF THE ART OF ILLUMINATION, for the Use of Beginners: with a Rudimentary Treatise on the Art, Practical Directions for its exercise, and Examples taken from Illuminated MSS., printed in Gold and Colours. By F. DELAMOTTE. New and Cheaper Edition. Small 4to, 6s. ornamental boards.

"The examples of ancient MSS. recommended to the student, which, with much good sense, the author chooses from collections accessible to all, are selected with judgment and knowledge, as well as taste."—*Athenæum.*

ORNAMENTAL ALPHABETS, Ancient and Mediæval, from the Eighth Century, with Numerals; including Gothic, Church-Text, large and small, German, Italian, Arabesque, Initials for Illumination, Monograms, Crosses, &c., for the use of Architectural and Engineering Draughtsmen, Missal Painters, Masons, Decorative Painters, Lithographers, Engravers, Carvers, &c. &c. Collected and Engraved by F. DELAMOTTE, and printed in Colours. New and Cheaper Edition. Royal 8vo, oblong, 2s. 6d. ornamental boards.

"For those who insert enamelled sentences round gilded chalices, who blazon shop legends over shop-doors, who letter church walls with pithy sentences from the Decalogue, this book will be useful."—*Athenæum.*

EXAMPLES OF MODERN ALPHABETS, Plain and Ornamental; including German, Old English, Saxon, Italic, Perspective, Greek, Hebrew, Court Hand, Engrossing, Tuscan, Riband, Gothic, Rustic, and Arabesque; with several Original Designs, and an Analysis of the Roman and Old English Alphabets, large and small, and Numerals, for the use of Draughtsmen, Surveyors, Masons, Decorative Painters, Lithographers, Engravers, Carvers, &c. Collected and Engraved by F. DELAMOTTE, and printed in Colours. New and Cheaper Edition. Royal 8vo, oblong, 2s. 6d. ornamental boards.

"There is comprised in it every possible shape into which the letters of the alphabet and numerals can be formed, and the talent which has been expended in the conception of the various plain and ornamental letters is wonderful."—*Standard.*

MEDIÆVAL ALPHABETS AND INITIALS FOR ILLUMINATORS. By F. G. DELAMOTTE. Containing 21 Plates and Illuminated Title, printed in Gold and Colours. With an Introduction by J. WILLIS BROOKS. Fourth and Cheaper Edition. Small 4to, 4s. ornamental boards.

"A volume in which the letters of the alphabet come forth glorified in gilding and all the colours of the prism interwoven and intertwined and intermingled."—*Sun.*

THE EMBROIDERER'S BOOK OF DESIGN. Containing Initials, Emblems, Cyphers, Monograms, Ornamental Borders, Ecclesiastical Devices, Mediæval and Modern Alphabets, and National Emblems. Collected by F. DELAMOTTE, and printed in Colours. Oblong royal 8vo, 1s. 6d. ornamental wrapper.

"The book will be of great assistance to ladies and young children who are endowed with the art of plying the needle in this most ornamental and useful pretty work."—*East Anglian Times.*

Wood Carving.

INSTRUCTIONS IN WOOD-CARVING, for Amateurs; with Hints on Design. By A LADY. With Ten Plates. New and Cheaper Edition. Crown 8vo, 2s. in emblematic wrapper.

"The handicraft of the wood-carver, so well as a book can impart it, may be learnt from 'A Lady's' publication."—*Athenæum.*

"The directions given are plain and easily understood."—*English Mechanic.*

Glass Painting.

GLASS STAINING AND THE ART OF PAINTING ON GLASS. From the German of Dr. GESSERT and EMANUEL OTTO FROMBERG. With an Appendix on THE ART OF ENAMELLING. 12mo, 2s. 6d. cloth limp.

Letter Painting.

THE ART OF LETTER PAINTING MADE EASY. By JAMES GREIG BADENOCH. With 12 full-page Engravings of Examples, 1s. 6d. cloth limp.

"The system is a simple ore, but quite original, and well worth the careful attention of letter painters. It can be easily mastered and remembered."—*Building News.*

CARPENTRY, TIMBER, etc.

Tredgold's Carpentry, Revised & Enlarged by Tarn.

THE ELEMENTARY PRINCIPLES OF CARPENTRY.
A Treatise on the Pressure and Equilibrium of Timber Framing, the Resistance of Timber, and the Construction of Floors, Arches, Bridges, Roofs, Uniting Iron and Stone with Timber, &c. To which is added an Essay on the Nature and Properties of Timber, &c., with Descriptions of the kinds of Wood used in Building; also numerous Tables of the Scantlings of Timber for different purposes, the Specific Gravities of Materials, &c. By THOMAS TREDGOLD, C.E. With an Appendix of Specimens of Various Roofs of Iron and Stone, Illustrated. Seventh Edition, thoroughly revised and considerably enlarged by E. WYNDHAM TARN, M.A., Author of "The Science of Building," &c. With 61 Plates, Portrait of the Author, and several Woodcuts. In one large vol., 4to, price £1 5s. cloth.

"Ought to be in every architect's and every builder's library."—*Builder.*
"A work whose monumental excellence must commend it wherever skilful carpentry is concerned. The author's principles are rather confirmed than impaired by time. The additional plates are of great intrinsic value."—*Building News.*

Woodworking Machinery.

WOODWORKING MACHINERY: Its Rise, Progress, and Construction. With Hints on the Management of Saw Mills and the Economical Conversion of Timber. Illustrated with Examples of Recent Designs by leading English, French, and American Engineers. By M. POWIS BALE, A.M.Inst.C.E., M.I.M.E. Large crown 8vo, 12s. 6d. cloth.

"Mr. Bale is evidently an expert on the subject and he has collected so much information that his book is all-sufficient for builders and others engaged in the conversion of timber."—*Architect.*
"The most comprehensive compendium of wood-working machinery we have seen. The author is a thorough master of his subject."—*Building News.*
"The appearance of this book at the present time will, we should think, give a considerable impetus to the onward march of the machinist engaged in the designing and manufacture of wood-working machines. It should be in the office of every wood-working factory."—*English Mechanic.*

Saw Mills.

SAW MILLS: Their Arrangement and Management, and the Economical Conversion of Timber. (A Companion Volume to "Woodworking Machinery.") By M. POWIS BALE. With numerous Illustrations. Crown 8vo, 10s. 6d. cloth.

"The *administration* of a large sawing establishment is discussed, and the subject examined from a financial standpoint. Hence the size, shape, order, and disposition of saw-mills and the like are gone into in detail, and the course of the timber is traced from its reception to its delivery in its converted state. We could not desire a more complete or practical treatise."—*Builder.*
"We highly recommend Mr. Bale's work to the attention and perusal of all those who are engaged in the art of wood conversion, or who are about building or remodelling saw-mills on improved principles."—*Building News.*

Carpentering.

THE CARPENTER'S NEW GUIDE; or, Book of Lines for Carpenters; comprising all the Elementary Principles essential for acquiring a knowledge of Carpentry. Founded on the late PETER NICHOLSON'S Standard Work. A New Edition, Revised by ARTHUR ASHPITEL, F.S.A. Together with Practical Rules on Drawing, by GEORGE PYNE. With 74 Plates, 4to, £1 1s. cloth.

Handrailing and Stairbuilding.

A PRACTICAL TREATISE ON HANDRAILING: Showing New and Simple Methods for Finding the Pitch of the Plank, Drawing the Moulds, Bevelling, Jointing-up, and Squaring the Wreath. By GEORGE COLLINGS. Second Edition, Revised and Enlarged, to which is added A TREATISE ON STAIRBUILDING. With Plates and Diagrams. 12mo, 2s. 6d. cloth limp. *[Just published.]*

"Will be found of practical utility in the execution of this difficult branch of joinery."—*Builder.*
"Almost every difficult phase of this somewhat intricate branch of joinery is elucidated by the aid of plates and explanatory letterpress."—*Furniture Gazette.*

Circular Work.

CIRCULAR WORK IN CARPENTRY AND JOINERY: A Practical Treatise on Circular Work of Single and Double Curvature. By GEORGE COLLINGS, Author of "A Practical Treatise on Handrailing." Illustrated with numerous Diagrams. Second Edition. 12mo, 2s. 6d. cloth limp.

"An excellent example of what a book of this kind should be. Cheap in price, clear in definition and practical in the examples selected."—*Builder.*

Timber Merchant's Companion.

THE TIMBER MERCHANT'S AND BUILDER'S COM-PANION. Containing New and Copious Tables of the Reduced Weight and Measurement of Deals and Battens, of all sizes, from One to a Thousand Pieces, and the relative Price that each size bears per Lineal Foot to any given Price per Petersburg Standard Hundred; the Price per Cube Foot of Square Timber to any given Price per Load of 50 Feet; the proportionate Value of Deals and Battens by the Standard, to Square Timber by the Load of 50 Feet; the readiest mode of ascertaining the Price of Scantling per Lineal Foot of any size, to any given Figure per Cube Foot, &c. &c.   By WILLIAM DOWSING. Fourth Edition, Revised and Corrected. Cr. 8vo, 3s. cl.

"Everything is as concise and clear as it can possibly be made. There can be no doubt that every timber merchant and builder ought to possess it."—*Hull Advertiser.*
" We are glad to see a fourth edition of these admirable tables, which for correctness and simplicity of arrangement leave nothing to be desired."—*Timber Trades Journal.*
"An exceedingly well-arranged, clear, and concise manual of tables for the use of all who buy or sell timber."—*Journal of Forestry.*

Practical Timber Merchant.

THE PRACTICAL TIMBER MERCHANT. Being a Guide for the use of Building Contractors, Surveyors, Builders, &c., comprising useful Tables for all purposes connected with the Timber Trade, Marks of Wood, Essay on the Strength of Timber, Remarks on the Growth of Timber, &c.   By W. RICHARDSON. Fcap. 8vo, 3s. 6d. cloth.

"This handy manual contains much valuable information for the use of timber merchants, builders, foresters, and all others connected with the growth, sale, and manufacture of timber.'—*Journal of Forestry.*

Timber Freight Book.

THE TIMBER MERCHANT'S, SAW MILLER'S, AND IMPORTER'S FREIGHT BOOK AND ASSISTANT. Comprising Rules, Tables, and Memoranda relating to the Timber Trade.   By WILLIAM RICHARDSON, Timber Broker; together with a Chapter on " SPEEDS OF SAW MILL MACHINERY," by M. POWIS BALE, M.I.M.E., &c.   12mo, 3s. 6d. cl. boards.

"A very useful manual of rules, tables, and memoranda relating to the timber trade. We recommend it as a compendium of calculation to all timber measurers and merchants, and as supplying a real want in the trade."—*Building News.*

Packing-Case Makers, Tables for.

PACKING-CASE TABLES; showing the number of Superficial Feet in Boxes or Packing-Cases, from six inches square and upwards. By W. RICHARDSON, Timber Broker. Second Edition. Oblong 4to, 3s. 6d. cl.

" Invaluable labour-saving tables."—*Ironmonger.*
"Will save much labour and calculation."—*Grocer.*

Superficial Measurement.

THE TRADESMAN'S GUIDE TO SUPERFICIAL MEA-SUREMENT. Tables calculated from 1 to 200 inches in length, by 1 to 108 inches in breadth. For the use of Architects, Surveyors, Engineers, Timber Merchants, Builders, &c.   By JAMES HAWKINGS.   Third Edition. Fcap., 3s. 6d. cloth.

" A useful collection of tables to facilitate rapid calculation of surfaces. The exact area of any surface of which the limits have been ascertained can be instantly determined. The book will be found of the greatest utility to all engaged in building operations."—*Scotsman.*
" These tables will be found of great assistance to all who require to make calculations in superficial measurement."—*English Mechanic.*

Forestry.

THE ELEMENTS OF FORESTRY. Designed to afford Information concerning the Planting and Care of Forest Trees for Ornament or Profit, with Suggestions upon the Creation and Care of Woodlands. By F. B. HOUGH.   Large crown 8vo, 10s. cloth.

Timber Importer's Guide.

THE TIMBER IMPORTER'S, TIMBER MERCHANT'S AND BUILDER'S STANDARD GUIDE. By RICHARD E. GRANDY. Comprising an Analysis of Deal Standards, Home and Foreign, with Comparative Values and Tabular Arrangements for fixing Nett Landed Cost on Baltic and North American Deals, including all intermediate Expenses, Freight, Insurance, &c. &c.   Together with copious Information for the Retailer and Builder.   Third Edition, Revised.   12mo, 2s. cloth limp.

"Everything it pretends to be: built up gradually, it leads one from a forest to a treenail, and throws in, as a makeweight, a host of material concerning bricks, columns, cisterns, &c."—*English Mechanic.*

MARINE ENGINEERING, NAVIGATION, etc.

Chain Cables.

CHAIN CABLES AND CHAINS. Comprising Sizes and Curves of Links, Studs, &c., Iron for Cables and Chains, Chain Cable and Chain Making, Forming and Welding Links, Strength of Cables and Chains, Certificates for Cables, Marking Cables, Prices of Chain Cables and Chains, Historical Notes, Acts of Parliament, Statutory Tests, Charges for Testing, List of Manufacturers of Cables, &c. &c. By THOMAS W. TRAILL, F.E.R.N., M. Inst. C.E., Engineer Surveyor in Chief, Board of Trade, Inspector of Chain Cable and Anchor Proving Establishments, and General Superintendent, Lloyd's Committee on Proving Establishments. With numerous Tables, Illustrations and Lithographic Drawings. Folio, £2 2s. cloth, bevelled boards.

"It contains a vast amount of valuable information. Nothing seems to be wanting to make it a complete and standard work of reference on the subject."—*Nautical Magazine.*

Marine Engineering.

MARINE ENGINES AND STEAM VESSELS (A Treatise on). By ROBERT MURRAY, C.E. Eighth Edition, thoroughly Revised, with considerable Additions by the Author and by GEORGE CARLISLE, C.E., Senior Surveyor to the Board of Trade at Liverpool. 12mo, 5s. cloth boards.

"Well adapted to give the young steamship engineer or marine engine and boiler maker a general introduction into his practical work."—*Mechanical World.*

"We feel sure that this thoroughly revised edition will continue to be as popular in the future as it has been in the past, as, for its size, it contains more useful information than any similar treatise."—*Industries.*

"As a compendious and useful guide to engineers of our mercantile and royal naval services we should say it cannot be surpassed."—*Building News.*

The information given is both sound and sensible, and well qualified to direct young seagoing hands on the straight road to the extra chief's certificate. Most useful to surveyors, inspectors, draughtsmen, and all young engineers who take an interest in their profession."—*Glasgow Herald.*

"An indispensable manual for the student of marine engineering."—*Liverpool Mercury.*

Pocket-Book for Naval Architects and Shipbuilders.

THE NAVAL ARCHITECT'S AND SHIPBUILDER'S POCKET-BOOK of Formulæ, Rules, and Tables, and MARINE ENGINEER'S AND SURVEYOR'S Handy Book of Reference. By CLEMENT MACKROW, Member of the Institution of Naval Architects, Naval Draughtsman. Fourth Edition, Revised. With numerous Diagrams, &c. Fcap., 12s. 6d. strongly bound in leather.

"Should be used by all who are engaged in the construction or design of vessels. . . . Will be found to contain the most useful tables and formulæ required by shipbuilders, carefully collected from the best authorities, and put together in a popular and simple form."—*Engineer.*

"The professional shipbuilder has now, in a convenient and accessible form, reliable data for solving many of the numerous problems that present themselves in the course of his work."—*Iron.*

"There is scarcely a subject on which a naval architect or shipbuilder can require to refresh his memory which will not be found within the covers of Mr. Mackrow's book."—*English Mechanic.*

Pocket-Book for Marine Engineers.

A POCKET-BOOK OF USEFUL TABLES AND FORMULÆ FOR MARINE ENGINEERS. By FRANK PROCTOR, A.I.N.A. Third Edition. Royal 32mo, leather, gilt edges, with strap, 4s.

"We recommend it to our readers as going far to supply a long-felt want."—*Naval Science.*

"A most useful companion to all marine engineers."—*United Service Gazette.*

Introduction to Marine Engineering.

ELEMENTARY ENGINEERING: A Manual for Young Marine Engineers and Apprentices. In the Form of Questions and Answers on Metals, Alloys, Strength of Materials, Construction and Management of Marine Engines and Boilers, Geometry, &c. &c. With an Appendix of Useful Tables. By JOHN SHERREN BREWER, Government Marine Surveyor, Hongkong. Small crown 8vo, 2s. cloth.

"Contains much valuable information for the class for whom it is intended, especially in the chapters on the management of boilers and engines."—*Nautical Magazine.*

"A useful introduction to the more elaborate text books."—*Scotsman.*

"To a student who has the requisite desire and resolve to attain a thorough knowledge, Mr. Brewer offers decidedly useful help."—*Athenæum.*

Navigation.

PRACTICAL NAVIGATION. Consisting of THE SAILOR'S SEA-BOOK, by JAMES GREENWOOD and W. H. ROSSER; together with the requisite Mathematical and Nautical Tables for the Working of the Problems, by HENRY LAW, C.E., and Professor J. R. YOUNG. Illustrated. 12mo, 7s, strongly half-bound.

MINING AND METALLURGY.

Metalliferous Mining in the United Kingdom.
BRITISH MINING: A Treatise on the History, Discovery, Practical Development, and Future Prospects of Metalliferous Mines in the United Kingdom. By ROBERT HUNT, F.R.S., Keeper of Mining Records; Editor of "Ure's Dictionary of Arts, Manufactures, and Mines," &c. Upwards of 950 pp., with 230 Illustrations. Second Edition, Revised. Super-royal 8vo, £2 2s. cloth.
"One of the most valuable works of reference of modern times. Mr. Hunt, as keeper of mining records of the United Kingdom, has had opportunities for such a task not enjoyed by anyone else, and has evidently made the most of them. . . . The language and style adopted are good, and the treatment of the various subjects laborious, conscientious, and scientific."—*Engineering.*
"The book is, in fact, a treasure-house of statistical information on mining subjects, and we know of no other work embodying so great a mass of matter of this kind. Were this the only merit of Mr. Hunt's volume, it would be sufficient to render it indispensable in this library of everyone interested in the development of the mining and metallurgical industries of this country."—*Athenæum.*
"A mass of information not elsewhere available, and of the greatest value to those who may be interested in our great mineral industries."—*Engineer.*
"A sound, business-like collection of interesting facts. . . . The amount of information Mr. Hunt has brought together is enormous. . . . The volume appears likely to convey more instruction upon the subject than any work hitherto published."—*Mining Journal.*

Colliery Management.
THE COLLIERY MANAGER'S HANDBOOK: A Comprehensive Treatise on the Laying-out and Working of Collieries, Designed as a Book of Reference for Colliery Managers, and for the Use of Coal-Mining Students preparing for First-class Certificates. By CALEB PAMELY, Mining Engineer and Surveyor; Member of the North of England Institute of Mining and Mechanical Engineers; and Member of the South Wales Institute of Mining Engineers. With nearly 500 Plans, Diagrams, and other Illustrations. Medium 8vo, about 600 pages. Price £1 5s. strongly bound.
[*Just ready.*

Coal and Iron.
THE COAL AND IRON INDUSTRIES OF THE UNITED KINGDOM. Comprising a Description of the Coal Fields, and of the Principal Seams of Coal, with Returns of their Produce and its Distribution, and Analyses of Special Varieties. Also an Account of the occurrence of Iron Ores in Veins or Seams; Analyses of each Variety; and a History of the Rise and Progress of Pig Iron Manufacture. By RICHARD MEADE, Assistant Keeper of Mining Records. With Maps. 8vo, £1 8s. cloth.
"The book is one which must find a place on the shelves of all interested in coal and iron production, and in the iron, steel, and other metallurgical industries."—*Engineer.*
"Of this book we may unreservedly say that it is the best of its class which we have ever met. . . . A book of reference which no one engaged in the iron or coal trades should omit from his library."—*Iron and Coal Trades Review.*

Prospecting for Gold and other Metals.
THE PROSPECTOR'S HANDBOOK: A Guide for the Prospector and Traveller in Search of Metal-Bearing or other Valuable Minerals. By J. W. ANDERSON, M.A. (Camb.), F.R.G.S., Author of "Fiji and New Caledonia." Fifth Edition, thoroughly Revised and Enlarged. Small crown 8vo, 3s. 6d. cloth.
"Will supply a much felt want, especially among Colonists, in whose way are so often thrown many mineralogical specimens the value of which it is difficult to determine."—*Engineer.*
"How to find commercial minerals, and how to identify them when they are found, are the leading points to which attention is directed. The author has managed to pack a much practical detail into his pages as would supply material for a book three times its size."—*Mining Journal.*

Mining Notes and Formulæ.
NOTES AND FORMULÆ FOR MINING STUDENTS. By JOHN HERMAN MERIVALE, M.A., Certificated Colliery Manager, Professor of Mining in the Durham College of Science, Newcastle-upon-Tyne. Third Edition, Revised and Enlarged. Small crown 8vo, 2s. 6d. cloth. [*Just published.*
"Invaluable to anyone who is working up for an examination on mining subjects."—*Coal and Iron Trades Review.*
"The author has done his work in an exceedingly creditable manner, and has produced a book that will be of service to students, and those who are practically engaged in mining operations."—*Engineer.*
"A vast amount of technical matter of the utmost value to mining engineers, and of considerable interest to students."—*Schoolmaster.*

Explosives.

A HANDBOOK ON MODERN EXPLOSIVES. Being a Practical Treatise on the Manufacture and Application of Dynamite, Gun-Cotton, Nitro-Glycerine and other Explosive Compounds. Including the Manufacture of Collodion-Cotton. By M. Eissler, Mining Engineer and Metallurgical Chemist, Author of "The Metallurgy of Gold," "The Metallurgy of Silver," &c. With about 100 Illustrations. Crown 8vo, 10s. 6d. cloth. [*Just published.*

Useful not only to the miner, but also to officers of both services to whom blasting and the use of explosives generally may at any time become a necessary auxiliary."—*Nature.*
"A veritable mine of information on the subject of explosives employed for military, mining and blasting purposes."—*Army and Navy Gazette.*
"The book is clearly written. Taken as a whole, we consider it an excellent little book and one that should be found of great service to miners and others who are engaged in work requiring the use of explosives."—*Athenæum.*

Gold, Metallurgy of.

THE METALLURGY OF GOLD: A Practical Treatise on the Metallurgical Treatment of Gold-bearing Ores. Including the Processes of Concentration and Chlorination, and the Assaying, Melting and Refining of Gold. By M. Eissler, Mining Engineer and Metallurgical Chemist, formerly Assistant Assayer of the U.S. Mint, San Francisco. Third Edition, Revised and greatly Enlarged. With 187 Illustrations. Crown 8vo, 12s. 6d. cloth.
[*Just published.*

"This book thoroughly deserves its title of a 'Practical Treatise.' The whole process of gold milling, from the breaking of the quartz to the assay of the bullion, is described in clear and orderly narrative and with much, but not too much, fulness of detail."—*Saturday Review.*
"The work is a storehouse of information and valuable data, and we strongly recommend it to all professional men engaged in the gold-mining industry."—*Mining Journal.*

Silver, Metallurgy of.

THE METALLURGY OF SILVER: A Practical Treatise on the Amalgamation, Roasting and Lixiviation of Silver Ores. Including the Assaying, Melting and Refining of Silver Bullion. By M. Eissler, Author of "The Metallurgy of Gold." With 124 Illustrations. Crown 8vo, 10s. 6d. cloth.

"A practical treatise, and a technical work which we are convinced will supply a long-felt want amongst practical men, and at the same time be of value to students and others indirectly connected with the industries."—*Mining Journal.*
"From first to last the book is thoroughly sound and reliable."—*Colliery Guardian.*
"For chemists, practical miners, assayers and investors alike, we do not know of any work on the subject so handy and yet so comprehensive."—*Glasgow Herald.*

Silver-Lead, Metallurgy of.

THE METALLURGY OF ARGENTIFEROUS LEAD ORES: A Practical Treatise on the Smelting of Silver-Lead Ores and the Refining of Lead Bullion. Illustrated with Plans and Sections of Smelting Furnaces and Plant in Europe and America. By M. Eissler, Author of "The Metallurgy of Gold," "The Metallurgy of Silver," &c. Cr. 8vo. [*In the press.*

Metalliferous Minerals and Mining.

TREATISE ON METALLIFEROUS MINERALS AND MINING. By D. C. Davies, F.G.S., Mining Engineer, &c., Author of "A Treatise on Slate and Slate Quarrying." Illustrated with numerous Wood Engravings. Fourth Edition, carefully Revised. Crown 8vo, 12s. 6d. cloth.

"Neither the practical miner nor the general reader interested in mines can have a better book for his companion and his guide."—*Mining Journal.*
"We are doing our readers a service in calling their attention to this valuable work "—
Mining World.
"A book that will not only be useful to the geologist, the practical miner, and the metallurgist, but also very interesting to the general public."—*Iron.*
"As a history of the present state of mining throughout the world this book has a real value, and it supplies an actual want."—*Athenæum.*

Earthy Minerals and Mining.

A TREATISE ON EARTHY & OTHER MINERALS AND MINING. By D. C. Davies, F.G.S. Uniform with, and forming a Companion Volume to, the same Author's "Metalliferous Minerals and Mining." With 76 Wood Engravings. Second Edition. Crown 8vo, 12s. 6d. cloth.

"We do not remember to have met with any English work on mining matters that contains the same amount of information packed in equally convenient form."—*Academy.*
"We should be inclined to rank it as among the very best of the handy technical and trade manuals which have recently appeared."—*British Quarterly Review.*

Mineral Surveying and Valuing.

THE MINERAL SURVEYOR AND VALUER'S COMPLETE
GUIDE, comprising a Treatise on Improved Mining Surveying and the Valua-
tion of Mining Properties, with New Traverse Tables. By WM. LINTERN,
Mining and Civil Engineer. Third Edition, with an Appendix on "Magnetic
and Angular Surveying," with Records of the Peculiarities of Needle Dis-
turbances. With Four Plates of Diagrams, Plans, &c. 12mo, 4s. cloth.
[Just published.
"Mr. Lintern's book forms a valuable and thoroughly trustworthy guide."—*Iron and Coal
Trades Review.*
"This new edition must be of the highest value to colliery surveyors, proprietors and mana-
gers."—*Colliery Guardian.*

Asbestos and its Uses.

ASBESTOS : Its Properties, Occurrence and Uses. With some
Account of the Mines of Italy and Canada. By ROBERT H. JONES. With
Eight Collotype Plates and other Illustrations. Crown 8vo, 12s. 6d. cloth.
[Just published.
"An interesting and invaluable work."—*Colliery Guardian.*
"We counsel our readers to get this exceedingly interesting work for themselves; they will
find in it much that is suggestive, and a great deal that is of immediate and practical usefulness."—
Builder.
"A valuable addition to the architect's and engineer's library."—*Building News.*

Underground Pumping Machinery.

MINE DRAINAGE. Being a Complete and Practical Treatise
on Direct-Acting Underground Steam Pumping Machinery, with a Description of a large number of the best known Engines, their General Utility and
the Special Sphere of their Action, the Mode of their Application, and
their merits compared with other forms of Pumping Machinery. By STEPHEN
MICHELL. 8vo, 15s. cloth.
"Will be highly esteemed by colliery owners and lessees, mining engineers, and students
generally who require to be acquainted with the best means of securing the drainage of mines. It
is a most valuable work, and stands almost alone in the literature of steam pumping machinery."—
Colliery Guardian.
"Much valuable information is given, so that the book is thoroughly worthy of an extensive
circulation amongst practical men and purchasers of machinery."—*Mining Journal.*

Mining Tools.

A MANUAL OF MINING TOOLS. For the Use of Mine
Managers, Agents, Students, &c. By WILLIAM MORGANS, Lecturer on Practical Mining at the Bristol School of Mines. 12mo, 2s. 6d. cloth limp.

ATLAS OF ENGRAVINGS to Illustrate the above, contain-
ing 235 Illustrations of Mining Tools, drawn to scale. 4to, 4s. 6d. cloth.
"Students in the science of mining, and overmen, captains, managers, and viewers may gain
practical knowledge and useful hints by the study of Mr. Morgans' manual."—*Colliery Guardian.*
"A valuable work, which will tend materially to improve our mining literature."—*Mining
Journal.*

Coal Mining.

COAL AND COAL MINING : A Rudimentary Treatise on. By
the late Sir WARINGTON W. SMYTH, M.A., F.R.S., &c., Chief Inspector of the
Mines of the Crown. Seventh Edition, Revised and Enlarged. With
numerous Illustrations. 12mo, 4s. cloth boards. *[Just published.*
"As an outline is given of every known coal-field in this and other countries, as well as of the
principal methods of working, the book will doubtless interest a very large number of readers."—
Mining Journal.

Subterraneous Surveying.

SUBTERRANEOUS SURVEYING, Elementary and Practical
Treatise on, with and without the Magnetic Needle. By THOMAS FENWICK,
Surveyor of Mines, and THOMAS BAKER, C.E. Illust. 12mo, 3s. cloth boards.

Granite Quarrying.

GRANITES AND OUR GRANITE INDUSTRIES. By
GEORGE F. HARRIS, F.G.S., Membre de la Société Belge de Géologie, Lec-
turer on Economic Geology at the Birkbeck Institution, &c. With Illustra-
tions. Crown 8vo, 2s. 6d. cloth.
"A clearly and well-written manual for persons engaged or interested in the granite industry."
—*Scotsman.*
"An interesting work, which will be deservedly esteemed."—*Colliery Guardian.*
"An exceedingly interesting and valuable monograph on a subject which has hitherto received
unaccountably little attention in the shape of systematic literary treatment."—*Scottish Leader.*

ELECTRICITY, ELECTRICAL ENGINEERING, etc.

Electrical Engineering.

THE ELECTRICAL ENGINEER'S POCKET-BOOK OF MODERN RULES, FORMULÆ, TABLES AND DATA. By H. R. KEMPE, M.Inst.E.E., A.M.Inst C.E., Technical Officer Postal Telegraphs, Author of " A Handbook of Electrical Testing," &c. With numerous Illustrations, royal 32mo, oblong, 5s. leather. [*Just published.*

" There is very little in the shape of formulæ or data which the electrician is likely to want in a hurry which cannot be found in its pages."—*Practical Engineer.*
" A very useful book of reference for daily use in practical electrical engineering and its various applications to the industries of the present day."—*Iron.*
" It is the best book of its kind."—*Electrical Engineer.*
" We l arranged and compact. The Electrical Engineer's Pocket-Book is a good one."—*Electrician.*
' Strongly recommended to those engaged in the various electrical industries."—*Electrical Review.*

Electric Lighting.

ELECTRIC LIGHT FITTING: A Handbook for Working Electrical Engineers, embodying Practical Notes on Installation Management. By JOHN W. URQUHART, Electrician, Author of " Electric Light," &c. With numerous Illustrations, crown 8vo, 5s. cloth. [*Just published.*

" This volume deals with what may be termed the mechanics of electric lighting, and is addressed to m n who are already engaged in the work or are training for it. The work traverses a great deal of ground, and may be read as a sequel to the same author's useful work on ' Electric Light.' "—*Electrician.*
" This is an attempt to state in the simplest language the precautions which should be adopted in instal ing the electric light, and to give information,for the guidance of those who have to run the plant when installed. The book is well worth the perusal of the workmen for whom it is written."—*Electrical Review.*
' Emin ntly practical and useful. . . . Ought to be in the hands of everyone in charge of an electric light plant."—*Electrical Engineer.*
" Altcgether Mr. Urquhart has succeeded in producing a really capital book, which we have no hesitation in recommending to the notice of working electricians and electrical engineers."—*Mechanical World.*

Electric Light.

ELECTRIC LIGHT : *Its Production and Use.* Embodying Plain Directions for the Treatment of Dynamo-Electric Machines, Bat'eries, Accumulators, and Electric Lamps. By J. W. URQUHART, C.E., Author of " Electric Light Fitting." " Electroplating," &c. Fourth Edition, carefully Revised, with Large Additions and 145 Illustrations. Crown 8vo, 7s. 6d. cloth. [*Just published.*

" The book is by far the best that we have yet met with on the subject."—*Athenæum.*
" It is the only work at present available which gives, in language intelligible for the most part to the ordinary reader, a general but concise history of the means which have been adopted up to the present time in producing the electric light."—*Metropolitan.*
" The book contains a general account of the means adopted in producing the electric light, not only as obtained from voltaic or galvanic batteries, but treats at length of the dynamo-electric machine in several of its forms."—*Colliery Guardian.*

Construction of Dynamos.

DYNAMO CONSTRUCTION : *A Practical Handbook for the Use of Engineer Constructors and Electricians in Charge.* With Examples of leading English, American and Continental Dynamos and Motors. By J. W. URQUHART, Author of " Electric Light," " Electric Light Fitting," &c. Crown 8vo. [*In the press.*

Text Book of Electricity.

THE STUDENT'S TEXT-BOOK OF ELECTRICITY. By HENRY M. NOAD, Ph.D., F.R.S., F.C.S. New Edition, carefully Revised. With an Introduction and Additional Chapters, by W. H. PREECE, M.I.C.E., Vice-President of the Society of Telegraph Engineers, &c. With 470 Illustrations. Crown 8vo, 12s. 6d. cloth.

" The original plan of this book has been carefully adhered to so as to make it a reflex of the existing state of electrical science, adapted for students. . . . Discovery seems to have progressed with marvellous strides ; nevertheless it has now apparently ceased, and practical applica tions have commenced their career ; and it is to give a faithful account of these that this fresh edition of Dr. Noad's valuable text-book is launched forth."—*Extract from Introduction by W. H. Preece, Esq.*
" We can recommend Dr. Noad's book for clear style, great range of subject, a good index and a plethora of woodcuts. Such collections as the present are indispensable."—*Athenæum.*
" An admirable text book for every student — beginner or advanced — of electricity."—*Engineering.*

Electric Lighting.
THE ELEMENTARY PRINCIPLES OF ELECTRIC LIGHT-
ING. By ALAN A. CAMPBELL SWINTON, Associate I.E.E. Second Edition,
Enlarged and Revised. With 16 Illustrations. Crown 8vo, 1s. 6d. cloth.
"Anyone who desires a short and thoroughly clear exposition of the elementary principles of
electric-lighting cannot do better than read this little work."—*Bradford Observer.*

Electricity.
A MANUAL OF ELECTRICITY: Including Galvanism, Mag-
netism, Dia-Magnetism, Electro-Dynamics, Magno-Electricity, and the Electric
Telegraph. By HENRY M. NOAD, Ph.D., F.R.S., F.C.S. Fourth Edition.
With 500 Woodcuts. 8vo, £1 4s. cloth.
"It is worthy of a place in the library of every public institution."—*Mining Journal.*

Dynamo Construction.
HOW TO MAKE A DYNAMO: A Practical Treatise for Amateurs.
Containing numerous Illustrations and Detailed Instructions for Construct-
ing a Small Dynamo, to Produce the Electric Light. By ALFRED CROFTS.
Third Edition, Revised and Enlarged. Crown 8vo, 2s. cloth.   [*Just published.*
"The instructions given in this unpretentious little book are sufficiently clear and explicit to
enable any amateur mechanic possessed of average skill and the usual tools to be found in an
amateur's workshop, to build a practical dynamo machine."—*Electrician.*

NATURAL SCIENCE, etc.

Pneumatics and Acoustics.
PNEUMATICS: including Acoustics and the Phenomena of Wind
Currents, for the Use of Beginners. By CHARLES TOMLINSON, F.R.S.,
F.C.S., &c. Fourth Edition, Enlarged. With numerous Illustrations.
12mo, 1s. 6d. cloth.
"Beginners in the study of this important application of science could not have a better manual."
—*Scotsman.*
"A valuable and suitable text-book for students of Acoustics and the Phenomena of Wind
Currents."—*Schoolmaster.*

Conchology.
A MANUAL OF THE MOLLUSCA: Being a Treatise on Recent
and Fossil Shells. By S. P. WOODWARD, A.L.S., F.G.S., late Assistant
Palæontologist in the British Museum. With an Appendix on Recent and
Fossil Conchological Discoveries, by RALPH TATE, A.L.S., F.G.S. Illustrated
by A. N. WATERHOUSE and JOSEPH WILSON LOWRY. With 23 Plates and
upwards of 300 Woodcuts. Reprint of Fourth Ed., 1880. Cr. 8vo, 7s. 6d. cl.
"A most valuable storehouse of conchological and geological information."—*Science Gossip.*

Geology.
RUDIMENTARY TREATISE ON GEOLOGY, PHYSICAL
AND HISTORICAL. Consisting of "Physical Geology," which sets forth
the leading Principles of the Science; and "Historical Geology," which
treats of the Mineral and Organic Conditions of the Earth at each successive
epoch, especial reference being made to the British Series of Rocks. By
RALPH TATE, A.L.S., F.G.S., &c. &c. With 250 Illustrations. 12mo, 5s.
cloth boards.
"The fulness of the matter has elevated the book into a manual. Its information is exhaustive
and well arranged."—*School Board Chronicle.*

Geology and Genesis.
THE TWIN RECORDS OF CREATION; or, Geology and
Genesis: their Perfect Harmony and Wonderful Concord. By GEORGE W.
VICTOR LE VAUX. Numerous Illustrations. Fcap. 8vo, 5s. cloth.
"A valuable contribution to the evidences of Revelation, and disposes very conclusively of the
arguments of those who would set God's Works against God's Word. No real difficulty is shirked,
and no sophistry is left unexposed."—*The Rock.*
"The remarkable peculiarity of this author is that he combines an unbounded admiration of
science with an unbounded admiration of the Written record. The two impulses are balanced to
a nicety; and the consequence is that difficulties, which to minds less evenly poised would be seri-
ous, find immediate solutions of the happiest kinds."—*London Review.*

Astronomy.
ASTRONOMY. By the late Rev. ROBERT MAIN, M.A., F.R.S.,
formerly Radcliffe Observer at Oxford. Third Edition, Revised and Cor-
rected to the present time, by WILLIAM THYNNE LYNN, B.A., F.R.A.S.,
formerly of the Royal Observatory, Greenwich. 12mo, 2s. cloth limp.
"A sound and simple treatise, very carefully edited, and a capital book for beginners."—
Knowledge.   [*tional Times.*
"Accurately brought down to the requirements of the present time by Mr. Lynn."—*Educa-*

DR. LARDNER'S COURSE OF NATURAL PHILOSOPHY.

THE HANDBOOK OF MECHANICS. Enlarged and almost re-written by BENJAMIN LOEWY, F.R.A.S. With 378 Illustrations. Post 8vo, 6s. cloth.

"The perspicuity of the original has been retained, and chapters which had become obsolete have been replaced by others of more modern character. The explanations throughout are studiously popular, and care has been taken to show the application of the various branches of physics to the industrial arts, and to the practical business of life."—*Mining Journal.*

"Mr. Loewy has carefully revised the book, and brought it up to modern requirements."—*Nature.*

"Natural philosophy has had few exponents more able or better skilled in the art of popularising the subject than Dr. Lardner; and Mr. Loewy is doing good service in fitting this treatise, and the others of the series, for use at the present time."—*Scotsman.*

THE HANDBOOK OF HYDROSTATICS AND PNEUMATICS. New Edition, Revised and Enlarged, by BENJAMIN LOEWY, F.R.A.S. With 236 Illustrations. Post 8vo, 5s. cloth.

"For those 'who desire to attain an accurate knowledge of physical science without the profound methods of mathematical investigation,' this work is not merely intended, but well adapted."—*Chemical News.*

"The volume before us has been carefully edited, augmented to nearly twice the bulk of the former edition, and all the most recent matter has been added. . . . It is a valuable text-book."—*Nature.*

"Candidates for pass examinations will find it, we think, specially suited to their requirements."—*English Mechanic.*

THE HANDBOOK OF HEAT. Edited and almost entirely re-written by BENJAMIN LOEWY, F.R.A.S., &c. 117 Illustrations. Post 8vo, 6s. cloth.

"The style is always clear and precise, and conveys instruction without leaving any cloudiness or lurking doubts behind."—*Engineering.*

"A most exhaustive book on the subject on which it treats, and is so arranged that it can be understood by all who desire to attain an accurate knowledge of physical science. . . . Mr. Loewy has included all the latest discoveries in the varied laws and effects of heat."—*Standard.*

"A complete and handy text-book for the use of students and general readers."—*English Mechanic.*

THE HANDBOOK OF OPTICS. By DIONYSIUS LARDNER, D.C.L., formerly Professor of Natural Philosophy and Astronomy in University College, London. New Edition. Edited by T. OLVER HARDING, B.A. Lond., of University College, London. With 298 Illustrations. Small 8vo, 448 pages, 5s. cloth.

"Written by one of the ablest English scientific writers, beautifully and elaborately illustrated."—*Mechanic's Magazine.*

THE HANDBOOK OF ELECTRICITY, MAGNETISM, AND ACOUSTICS. By Dr. LARDNER. Ninth Thousand. Edit. by GEORGE CAREY FOSTER, B.A., F.C.S. With 400 Illustrations. Small 8vo, 5s. cloth.

"The book could not have been entrusted to anyone better calculated to preserve the terse and lucid style of Lardner, while correcting his errors and bringing up his work to the present state of scientific knowledge."—*Popular Science Review.*

THE HANDBOOK OF ASTRONOMY. Forming a Companion to the "Handbook of Natural Philosophy." By DIONYSIUS LARDNER, D.C.L., formerly Professor of Natural Philosophy and Astronomy in University College, London. Fourth Edition. Revised and Edited by EDWIN DUNKIN, F.R.A.S., Royal Observatory, Greenwich. With 38 Plates and upwards of 100 Woodcuts. In One Vol., small 8vo, 550 pages, 9s. 6d. cloth.

"Probably no other book contains the same amount of information in so compendious and well-arranged a form—certainly none at the price at which this is offered to the public."—*Athenæum.*

"We can do no other than pronounce this work a most valuable manual of astronomy, and we strongly recommend it to all who wish to acquire a general—but at the same time correct—acquaintance with this sublime science."—*Quarterly Journal of Science.*

"One of the most deservedly popular books on the subject . . . We would recommend not only the student of the elementary principles of the science, but he who aims at mastering the higher and mathematical branches of astronomy, not to be without this work beside him."—*Practical Magazine.*

Dr. Lardner's Electric Telegraph.

THE ELECTRIC TELEGRAPH. By Dr. LARDNER. Revised and Re-written by E. B. BRIGHT, F.R.A.S. 140 Illustrations. Small 8vo, 2s. 6d. cloth.

"One of the most readable books extant on the Electric Telegraph."—*English Mechanic.*

DR. LARDNER'S MUSEUM OF SCIENCE AND ART.

THE MUSEUM OF SCIENCE AND ART. Edited by
DIONYSIUS LARDNER, D.C.L., formerly Professor of Natural Philosophy and
Astronomy in University College, London. With upwards of 1,200 Engravings on Wood. In 6 Double Volumes, £1 1s., in a new and elegant cloth binding; or handsomely bound in half-morocco, 31s. 6d.

*** OPINIONS OF THE PRESS.

"This series, besides affording popular but sound instruction on scientific subjects, with which
the humblest man in the country ought to be acquainted, also undertakes that teaching of 'Common Things' which every well-wisher of his kind is anxious to promote. Many thousand copies of
this serviceable publication have been printed, in the belief and hope that the desire for instruction
and improvement widely prevails; and we have no fear that such enlightened faith will meet with
disappointment."—*Times.*

"A cheap and interesting publication, alike informing and attractive. The papers combine
subjects of importance and great scientific knowledge, considerable inductive powers, and a
popular style of treatment."—*Spectator.*

"The 'Museum of Science and Art' is the most valuable contribution that has ever been
made to the Scientific Instruction of every class of society."—Sir DAVID BREWSTER, in the
North British Review.

"Whether we consider the liberality and beauty of the illustrations, the charm of the writing,
or the durable interest of the matter, we must express our belief that there is hardly to be found
among the new books one that would be welcomed by people of so many ages and classes as a
valuable present."—*Examiner.*

*** *Separate books formed from the above, suitable for Workmen's Libraries,*
Science Classes, etc.

Common Things Explained. Containing Air, Earth, Fire, Water, Time,
Man, the Eye, Locomotion, Colour, Clocks and Watches, &c. 233 Illustrations, cloth gilt, 5s.

The Microscope. Containing Optical Images, Magnifying Glasses, Origin
and Description of the Microscope, Microscopic Objects, the Solar Microscope, Microscopic Drawing and Engraving, &c. 147 Illustrations, cloth
gilt, 2s.

Popular Geology. Containing Earthquakes and Volcanoes, the Crust of
the Earth, &c. 201 Illustrations, cloth gilt, 2s. 6d.

Popular Physics. Containing Magnitude and Minuteness, the Atmosphere, Meteoric Stones, Popular Fallacies, Weather Prognostics, the
Thermometer, the Barometer, Sound, &c. 85 Illustrations, cloth gilt, 2s. 6d.

Steam and its Uses. Including the Steam Engine, the Locomotive, and
Steam Navigation   89 Illustrations, cloth gilt, 2s.

Popular Astronomy. Containing How to observe the Heavens—The
Earth, Sun, Moon, Planets, Light, Comets, Eclipses, Astronomical Influ-
ences, &c. 182 Illustrations, 4s. 6d.

The Bee and White Ants: Their Manners and Habits. With Illustrations of Animal Instinct and Intelligence. 135 Illustrations, cloth gilt, 2s.

The Electric Telegraph Popularized. To render intelligible to all who
can Read, irrespective of any previous Scientific Acquirements, the various
forms of Telegraphy in Actual Operation.   100 Illustrations, cloth gilt,
1s. 6d.

Dr. Lardner's School Handbooks.

NATURAL PHILOSOPHY FOR SCHOOLS. By Dr. LARDNER.
328 Illustrations. Sixth Edition. One Vol., 3s. 6d. cloth.

"A very convenient class-book for junior students in private schools. It is intended to convey,
in clear and precise terms, general notions of all the principal divisions of Physical Science."—
British Quarterly Review.

ANIMAL PHYSIOLOGY FOR SCHOOLS. By Dr. LARDNER.
With 190 Illustrations. Second Edition. One Vol., 3s. 6d. cloth.

"Clearly written, well arranged, and excellently illustrated."—*Gardener's Chronicle.*

COUNTING-HOUSE WORK, TABLES, etc.

Accounts for Manufacturers.

FACTORY ACCOUNTS: Their Principles and Practice. A Handbook for Accountants and Manufacturers, with Appendices on the Nomenclature of Machine Details; the Income Tax Acts; the Rating of Factories; Fire and Boiler Insurance; the Factory and Workshop Acts, &c., including also a Glossary of Terms and a large number of Specimen Rulings. By EMILE GARCKE and J. M. FELLS. Third Edition. Demy 8vo, 250 pages, price 6s. strongly bound.

"A very interesting description of the requirements of Factory Accounts. . . . the principle of assimilating the Factory Accounts to the general commercial books is one which we thoroughly agree with."—*Accountants' Journal*.

"Characterised by extreme thoroughness. There are few owners of Factories who would not derive great benefit from the perusal of this most admirable work."—*Local Government Chronicle*.

Foreign Commercial Correspondence.

THE FOREIGN COMMERCIAL CORRESPONDENT: Being Aids to Commercial Correspondence in Five Languages—English, French, German, Italian and Spanish. By CONRAD E. BAKER. Second Edition, Revised. Crown 8vo, 3s. 6d. cloth.

"Whoever wishes to correspond in all the languages mentioned by Mr. Baker cannot do better than study this work, the materials of which are excellent and conveniently arranged. They consist not of entire specimen letters, but what are far more useful—short passages, sentences, or phrases expressing the same general idea in various forms."—*Athenæum*.

"A careful examination has convinced us that it is unusually complete, well arranged and reliable. The book is a thoroughly good one."—*Schoolmaster*.

Intuitive Calculations.

THE COMPENDIOUS CALCULATOR; or, Easy and Concise Methods of Performing the various Arithmetical Operations required in Commercial and Business Transactions, together with Useful Tables. By DANIEL O'GORMAN. Corrected and Extended by J. R. YOUNG, formerly Professor of Mathematics at Belfast College. Twenty-seventh Edition, carefully Revised by C. NORRIS. Fcap. 8vo, 2s. 6d. cloth limp; or, 3s. 6d. strongly half-bound in leather.

"It would be difficult to exaggerate the usefulness of a book like this to everyone engaged in commerce or manufacturing industry. It is crammed full of rules and formulæ for shortening and employing calculations."—*Knowledge*.

"Supplies special and rapid methods for all kinds of calculations. Of great utility to persons engaged in any kind of commercial transactions."—*Scotsman*.

Modern Metrical Units and Systems.

MODERN METROLOGY: A Manual of the Metrical Units and Systems of the Present Century. With an Appendix containing a proposed English System. By LOWIS D'A. JACKSON, A.M.Inst.C.E., Author of "Aid to Survey Practice," &c. Large crown 8vo, 12s. 6d. cloth.

"The author has brought together much valuable and interesting information. . . . We cannot but recommend the work to the consideration of all interested in the practical reform of our weights and measures."—*Nature*.

"For exhaustive tables of equivalent weights and measures of all sorts, and for clear demonstrations of the effects of the various systems that have been proposed or adopted, Mr. Jackson's treatise is without a rival."—*Academy*.

The Metric System and the British Standards.

A SERIES OF METRIC TABLES, in which the British Standard Measures and Weights are compared with those of the Metric System at present in Use on the Continent. By C. H. DOWLING, C.E. 8vo, 10s. 6d. strongly bound.

"Their accuracy has been certified by Professor Airy, the Astronomer-Royal."—*Builder*.

"Mr. Dowling's Tables are well put together as a ready-reckoner for the conversion of one system into the other."—*Athenæum*.

Iron and Metal Trades' Calculator.

THE IRON AND METAL TRADES' COMPANION. For expeditiously ascertaining the Value of any Goods bought or sold by Weight, from 1s. per cwt. to 112s. per cwt., and from one farthing per pound to one shilling per pound. Each Table extends from one pound to 100 tons. To which are appended Rules on Decimals, Square and Cube Root, Mensuration of Superficies and Solids, &c.; also Tables of Weights of Materials, and other Useful Memoranda. By THOS. DOWNIE. Strongly bound in leather, 396 pp., 9s.

"A most useful set of tables, and will supply a want, for nothing like them before existed."—*Building News*.

"Although specially adapted to the iron and metal trades, the tables will be found useful in every other business in which merchandise is bought and sold by weight."—*Railway News*

Calculator for Numbers and Weights Combined.

**THE NUMBER, WEIGHT AND FRACTIONAL CALCU-
LATOR.** Containing upwards of 250,000 Separate Calculations, showing at
a glance the value at 422 different rates, ranging from $\frac{1}{16}$th of a Penny to
20s. each, or per cwt., and £20 per ton, of any number of articles consecu-
tively, from 1 to 470.—Any number of cwts., qrs., and lbs., from 1 cwt. to 470
cwts.—Any number of tons, cwts., qrs., and lbs., from 1 to 1,000 tons. By
WILLIAM CHADWICK, Public Accountant. Third Edition, Revised and Improved. 8vo, price 18s., strongly bound for Office wear and tear. [*Just published.*

*** *This work is specially adapted for the Apportionment of Mileage Charges
for Railway Traffic.*

☞ *This comprehensive and entirely unique and original Calculator is adapted
for the use of Accountants and Auditors, Railway Companies, Canal Companies,
Shippers, Shipping Agents, General Carriers, etc.*

*Ironfounders, Brassfounders, Metal Merchants, Iron Manufacturers,Ironmongers,
Engineers, Machinists, Boiler Makers, Millwrights, Roofing, Bridge and Girder
Makers, Colliery Proprietors, etc.*

*Timber Merchants, Builders, Contractors, Architects, Surveyors, Auctioneers
Valuers, Brokers, Mill Owners and Manufacturers, Mill Furnishers, Merchants and
General Wholesale Tradesmen.*

*** OPINIONS OF THE PRESS.

"The book contains the answers to questions, and not simply a set of ingenious puzzle
methods of arriving at results. It is as easy of reference for any answer or any number of answers
as a dictionary, and the references are even more quickly made. For making up accounts or estimates, the book must prove invaluable to all who have any considerable quantity of calculations
involving price and measure in any combination to do."—*Engineer.*

"The most perfect work of the kind yet prepared."—*Glasgow Herald.*

Comprehensive Weight Calculator.

THE WEIGHT CALCULATOR. Being a Series of Tables
upon a New and Comprehensive Plan, exhibiting at One Reference the exact
Value of any Weight from 1 lb. to 15 tons, at 300 Progressive Rates, from 1d.
to 168s. per cwt., and containing 186,000 Direct Answers, which, with their
Combinations, consisting of a single addition (mostly to be performed at
sight), will afford an aggregate of 10,266,000 Answers; the whole being calculated and designed to ensure correctness and promote despatch. By HENRY
HARBEN, Accountant. Fourth Edition, carefully Corrected. Royal 8vo,
strongly half-bound, £1 5s.

"A practical and useful work of reference for men of business generally ; it is the best of the
kind we have seen.'—*Ironmonger.*

"Of priceless value to business men. It is a necessary book in all mercantile offices."—*Sheffield Independent.*

Comprehensive Discount Guide.

THE DISCOUNT GUIDE. Comprising several Series of
Tables for the use of Merchants, Manufacturers, Ironmongers, and others,
by which may be ascertained the exact Profit arising from any mode of using
Discounts, either in the Purchase or Sale of Goods, and the method of either
Altering a Rate of Discount or Advancing a Price, so as to produce, by one
operation, a sum that will realise any required profit after allowing one or
more Discounts : to which are added Tables of Profit or Advance from 1¼ to
90 per cent., Tables of Discount from 1½ to 98¾ per cent., and Tables of Commission, &c., from ⅛ to 10 per cent. By HENRY HARBEN, Accountant, Author
of "The Weight Calculator." New Edition, carefully Revised and Corrected.
Demy 8vo, 544 pp. half-bound, £1 5s.

"A book such as this can only be appreciated by business men, to whom the saving of time
means saving of money. We have the high authority of Professor J. R. Young that the tables
throughout the work are constructed upon strictly accurate principles. The work is a model
of typographical clearness, and must prove of great value to merchants, manufacturers, and
general traders."—*British Trade Journal.*

Iron Shipbuilders' and Merchants' Weight Tables.

IRON-PLATE WEIGHT TABLES: For Iron Shipbuilders,
Engineers and Iron Merchants. Containing the Calculated Weights of upwards of 150,000 different sizes of Iron Plates, from 1 foot by 6 in. by ¼ in. to
10 feet by 5 feet by 1 in. Worked out on the basis of 40 lbs. to the square
foot of Iron of 1 inch in thickness. Carefully compiled and thoroughly Revised by H. BURLINSON and W. H. SIMPSON. Oblong 4to, 25s. half-bound.

"This work will be found of great utility. The authors have had much practical experience
of what is wanting in making estimates; and the use of the book will save much time in making
elaborate calculations."—*English Mechanic*

INDUSTRIAL AND USEFUL ARTS.

Soap-making.
THE ART OF SOAP-MAKING : A Practical Handbook of the *Manufacture of Hard and Soft Soaps, Toilet Soaps, etc.* Including many New Processes, and a Chapter on the Recovery of Glycerine from Waste Leys. By ALEXANDER WATT, Author of "Electro-Metallurgy Practically Treated," &c. With numerous Illustrations. Fourth Edition, Revised and Enlarged. Crown 8vo, 7s. 6d. cloth. [*Just published.*

"The work will prove very useful, not merely to the technological student, but to the practical soap-boiler who wishes to understand the theory of his art."—*Chemical News.*

"Mr. Watt's book is a thoroughly practical treatise on an art which has almost no literature in our language. We congratulate the author on the success of his endeavour to fill a void in English technical literature."—*Nature.*

Paper Making.
THE ART OF PAPER MAKING : A Practical Handbook of the *Manufacture of Paper from Rags, Esparto, Straw and other Fibrous Materials,* Including the Manufacture of Pulp from Wood Fibre, with a Description of the Machinery and Appliances used. To which are added Details of Processes for Recovering Soda from Waste Liquors. By ALEXANDER WATT. With Illustrations. Crown 8vo, 7s. 6d. cloth. [*Just published.*

" This book is succinct, lucid, thoroughly practical, and includes everything of interest to the modern paper maker. It is the latest, most practical and most complete work on the paper-making art before the British public."—*Paper Record.*

' It may be regarded as the standard work on the subject. The book is full of valuable information. The 'Art of Paper-making,' is in every respect a model of a text-book, either for a technical class or for the private student."—*Paper and Printing Trades Journal.*

" Admirably adapted for general as well as ordinary technical reference, and as a handbook for students in technical education may be warmly commended."—*The Paper Maker's Monthly Journal.*

Leather Manufacture.
THE ART OF LEATHER MANUFACTURE. Being a Practical Handbook, in which the Operations of Tanning, Currying, and Leather Dressing are fully Described, the Principles of Tanning Explained and many Recent Processes introduced. By ALEXANDER WATT, Author of " Soap-Making," &c. With numerous Illustrations. Second Edition. Crown 8vo, 9s. cloth.

"A sound, comprehensive treatise on tanning and its accessories. This book is an eminently valuable production, which redounds to the credit of both author and publishers."—*Chemical Review.*

"This volume is technical without being tedious, comprehensive and complete without being prosy, and it bears on every page the impress of a master hand. We have never come across a better trade treatise, nor one that so thoroughly supplied an absolute want."—*Shoe and Leather Trades' Chronicle.*

Boot and Shoe Making.
THE ART OF BOOT AND SHOE-MAKING. A Practical Handbook, including Measurement, Last-Fitting, Cutting-Out, Closing and Making, with a Description of the most approved Machinery employed. By JOHN B. LENO, late Editor of *St. Crispin,* and *The Boot and Shoe-Maker.* With numerous Illustrations. Third Edition. 12mo, 2s. cloth limp.

"This excellent treatise is by far the best work ever written on the subject. A new work, embracing all modern improvements, was much wanted. This want is now satisfied. The chapter on clicking, which shows how waste may be prevented, will save fifty times the price of the book." —*Scottish Leather Trader.*

Dentistry.
MECHANICAL DENTISTRY : A Practical Treatise on the *Construction of the various kinds of Artificial Dentures.* Comprising also Useful Formulæ, Tables and Receipts for Gold Plate, Clasps, Solders, &c. &c. By CHARLES HUNTER. Third Edition, Revised. With upwards of 100 Wood Engravings. Crown 8vo, 3s. 6d. cloth.

" The work is very practical."—*Monthly Review of Dental Surgery.*

" We can strongly recommend Mr. Hunter's treatise to all students preparing for the profession of dentistry, as well as to every mechanical dentist."—*Dublin Journal of Medical Science.*

Wood Engraving.
WOOD ENGRAVING: A Practical and Easy Introduction to the *Study of the Art.* By WILLIAM NORMAN BROWN. Second Edition. With numerous Illustrations. 12mo, 1s. 6d. cloth limp.

" The book is clear and complete, and will be useful to anyone wanting to understand the first elements of the beautiful art of wood engraving."—*Graphic.*

HANDYBOOKS FOR HANDICRAFTS. By PAUL N. HASLUCK.

Metal Turning.
THE METAL TURNER'S HANDBOOK. *A Practical Manual for Workers at the Foot-Lathe:* Embracing Information on the Tools, Appliances and Processes employe i in Metal Turning. By PAUL N. HASLUCK, Author of " Lathe-Work." With upwards of One Hundred Illustrations. Second Edition, Revised. Crown 8vo, 2s. cloth.
"Clearly and concisely written, excellent in every way."—*Mechanical World.*

Wood Turning.
THE WOOD TURNER'S HANDYBOOK. *A Practical Manual for Workers at the Lathe:* Embracing Information on the Tools, Appliances and Processes Employed in Wood Turning. By PAUL N. HASLUCK. With upwards of One Hundred Illustrations. Crown 8vo, 2s. cloth.
"We recommend the book to young turners and amateurs. A multitude of workmen have hitherto sought in vain for a manual of this special industry."—*Mechanical World.*

WOOD AND METAL TURNING. By P. N. HASLUCK. (Being the Two preceding Vols. bound together.) 300 pp, with upwards of 200 Illustrations, crown 8vo, 3s. 6d. cloth.

Watch Repairing.
THE WATCH JOBBER'S HANDYBOOK. *A Practical Manual on Cleaning, Repairing and Adjusting.* Embracing Information on the Tools, Materials, Appliances and Processes Employed in Watchwork. By PAUL N. HASLUCK. With upwards of One Hundred Illustrations. Cr. 8vo, 2s. cloth.
"All young persons connected with the trade should acquire and study this excellent, and at the same time, inexpensive work."—*Clerkenwell Chronicle.*

Clock Repairing.
THE CLOCK JOBBER'S HANDYBOOK: *A Practical Manual on Cleaning, Repairing and Adjusting.* Embracing Information on the Tools, Materials, Appliances and Processes Employed in Clockwork. By PAUL N. HASLUCK. With upwards of 100 Illustrations. Cr. 8vo, 2s. cloth.
"Of inestimable service to those commencing the trade."—*Coventry Standard.*

WATCH AND CLOCK JOBBING. By P. N. HASLUCK. (Being the Two preceding Vols. bound together.) 320 pp., with upwards of 200 Illustrations, crown 8vo, 3s. 6d. cloth.

Pattern Making.
THE PATTERN MAKER'S HANDYBOOK. A Practical Manual, embracing Information on the Tools, Materials and Appliances employed in Constructing Patterns for Founders. By PAUL N. HASLUCK. With One Hundred Illustrations. Crown 8vo, 2s. cloth.
"This handy volume contains sound information of considerable value to students and artificers."—*Hardware Trades Journal.*

Mechanical Manipulation.
THE MECHANIC'S WORKSHOP HANDYBOOK. *A Practical Manual on Mechanical Manipulation.* Embracing Information on various Handicraft Processes, with Useful Notes and Miscellaneous Memoranda. By PAUL N. HASLUCK. Crown 8vo, 2s. cloth.
"It is a book which should be found in every workshop, as it is one which will be continually referred to for a very great amount of standard information."—*Saturday Review.*

Model Engineering.
THE MODEL ENGINEER'S HANDYBOOK: *A Practical Manual on Model Steam Engines.* Embracing Information on the Tools, Materials and Processes Employed in their Construction. By PAUL N. HASLUCK. With upwards of 100 Illustrations. Crown 8vo, 2s. cloth.
"By carefully going through the work, amateurs may pick up an excellent notion of the construction of full-sized steam engines."—*Telegraphic Journal.*

Cabinet Making.
THE CABINET WORKER'S HANDYBOOK: A Practical Manual, embracing Information on the Tools, Materials, Appliances and Processes employed in Cabinet Work. By PAUL N. HASLUCK, Author of " Lathe Work," &c. With upwards of 100 Illustrations. Crown 8vo, 2s. cloth. *[Just published.*
"Thoroughly practical throughout. The amateur worker in wood will find it most useful."—*Glasgow Herald.*

Electrolysis of Gold, Silver, Copper, etc.

ELECTRO-DEPOSITION : A Practical Treatise on the Electrolysis of Gold, Silver, Copper, Nickel, and other Metals and Alloys. With descriptions of Voltaic Batteries, Magneto and Dynamo-Electric Machines, Thermopiles, and of the Materials and Processes used in every Department of the Art, and several Chapters on Electro-Metallurgy. By ALEXANDER WATT. Third Edition, Revised and Corrected. Crown 8vo, 9s. cloth.

"Eminently a book for the practical worker in electro-deposition. It contains practical descriptions of methods, processes and materials as actually pursued and used in the workshop." —*Engineer.*

Electro-Metallurgy.

ELECTRO-METALLURGY ; Practically Treated. By ALEXANDER WATT. Author of "Electro-Deposition," &c. Ninth Edition, Enlarged and Revised, with Additional Illustrations, and including the most recent Processes. 12mo, 4s. cloth boards.

"From this book both amateur and artisan may learn everything necessary for the successful prosecution of electroplating."—*Iron.*

Electroplating.

ELECTROPLATING : A Practical Handbook on the Deposition of Copper, Silver, Nickel, Gold, Aluminium, Brass, Platinum, &c. &c. With Descriptions of the Chemicals, Materials, 'Batteries and Dynamo Machines used in the Art. By J. W. URQUHART, C.E. Second Edition, with Additions. Numerous Illustrations. Crown 8vo, 5s. cloth.

" An excellent practical manual."—*Engineering.*
" An excellent work, giving the newest information."—*Horological Journal.*

Electrotyping.

ELECTROTYPING : The Reproduction and Multiplication of Printing Surfaces and Works of Art by the Electro-deposition of Metals. By J. W. URQUHART, C.E. Crown 8vo, 5s. cloth.

'The book is thoroughly practical. The reader is, therefore, conducted through the leading aws of electricity, then through the metals used by electrotypers, the apparatus, and the depositing processes, up to the final preparation of the work."—*Art Journal.*

Horology.

A TREATISE ON MODERN HOROLOGY, in Theory and Practice. Translated from the French of CLAUDIUS SAUNIER, by JULIEN TRIPPLIN, F.R.A.S., and EDWARD RIGG, M.A., Assayer in the Royal Mint. 78 Woodcuts and 22 Coloured Plates. Second Edition. Royal 8vo, £2 2s. cloth ; £2 10s. half-calf.

" There is no horological work in the English language at all to be compared to this production of M. Saunier's for clearness and completeness. It is alike good as a guide for the student and as a reference for the experienced horologist and skilled workman."—*Horological Journal.*
" The latest, the most complete, and the most reliable of those literary productions to which continental watchmakers are indebted for the mechanical superiority over their English brethren —in fact, the Book of Books, is M. Saunier's 'Treatise.'"—*Watchmaker, Jeweller and Silversmith.*

Watchmaking.

THE WATCHMAKER'S HANDBOOK. A Workshop Companion for those engaged in Watchmaking and the Allied Mechanical Arts. From the French of CLAUDIUS SAUNIER. Enlarged by JULIEN TRIPPLIN, F.R.AS., and EDWARD RIGG, M.A., Assayer in the Royal Mint. Woodcuts and Copper Plates. Third Edition, Revised. Crown 8vo, 9s. cloth.

" Each part is truly a treatise in itself. The arrangement is good and the language is clear and concise. It is an admirable guide for the young watchmaker."—*Engineering.*
" It is impossible to speak too highly of its excellence. It fulfils every requirement in a handbook intended for the use of a workman."—*Watch and Clockmaker.*
" This book contains an immense number of practical details bearing on the daily occupation of a watchmaker."—*Watchmaker and Metalworker* (Chicago).

Goldsmiths' Work.

THE GOLDSMITH'S HANDBOOK. By GEORGE E. GEE, Jeweller, &c. Third Edition, considerably Enlarged. 12mo, 3s. 6d. cl. bds.
" A good, sound educator, and will be accepted as an authority."—*Horological Journal.*

Silversmiths' Work.

THE SILVERSMITH'S HANDBOOK. By GEORGE E. GEE, Jeweller, &c. Second Edition, Revised, with numerous Illustrations. 12mo. 3s. 6d. cloth boards.

"Workers in the trade will speedily discover its merits when they sit down to study it."—*English Mechanic.*

** The above two works together, strongly half-bound, price 7s.

D

Bread and Biscuit Baking.

THE BREAD AND BISCUIT BAKER'S AND SUGAR-BOILER'S ASSISTANT. Including a large variety of Modern Recipes. With Remarks on the Art of Bread-making. By ROBERT WELLS, Practical Baker. Second Edition, with Additional Recipes. Crown 8vo, 2s. cloth.

[Just published.

" A large number of wrinkles for the ordinary cook, as well as the baker."—*Saturday Review.*

Confectionery.

THE PASTRYCOOK AND CONFECTIONER'S GUIDE. For Hotels, Restaurants and the Trade in general, adapted also for Family Use. By ROBERT WELLS, Author of " The Bread and Biscuit Baker's and Sugar Boiler's Assistant." Crown 8vo, 2s. cloth. [Just published.

" We cannot speak too highly of this really excellent work. In these days of keen competition our readers cannot do better than purchase this book."—*Bakers' Times.*

Ornamental Confectionery.

ORNAMENTAL CONFECTIONERY: A Guide for Bakers, Confectioners and Pastrycooks; including a variety of Modern Recipes, and Remarks on Decorative and Coloured Work. With 129 Original Designs. By ROBERT WELLS. Crown 8vo, 5s. cloth.

" A valuable work, practical, and should be in the hands of every baker and confectioner. The illustrative designs are alone worth treble the amount charged for the whole work."—*Bakers' Times.*

Flour Confectionery.

THE MODERN FLOUR CONFECTIONER. Wholesale and Retail. Containing a large Collection of Recipes for Cheap Cakes, Biscuits, &c. With Remarks on the Ingredients used in their Manufacture, &c. By R. WELLS, Author of " Ornamental Confectionery," " The Bread and Biscuit Baker," " The Pastrycook's Guide," &c. Crown 8vo, 2s. cloth.

[Just published.

Laundry Work.

LAUNDRY MANAGEMENT. A Handbook for Use in Private and Public Laundries, Including Descriptive Accounts of Modern Machinery and Appliances for Laundry Work. By the EDITOR of " The Laundry Journal." With numerous Illustrations. Crown 8vo, 2s. 6d. cloth.

CHEMICAL MANUFACTURES & COMMERCE.

Alkali Trade, Manufacture of Sulphuric Acid, etc.

A MANUAL OF THE ALKALI TRADE, including the Manufacture of Sulphuric Acid, Sulphate of Soda, and Bleaching Powder. By JOHN LOMAS. 390 pages. With 232 Illustrations and Working Drawings. Second Edition. Royal 8vo, £1 10s. cloth.

" This book is written by a manufacturer for manufacturers. The working details of the most approved forms of apparatus are given, and these are accompanied by no less than 232 wood engravings, all of which may be used for the purposes of construction. Every step in the manufacture is very fully described in this manual, and each improvement explained."—*Athenæum.*

The Blowpipe.

THE BLOWPIPE IN CHEMISTRY, MINERALOGY, AND GEOLOGY. Containing all known Methods of Anhydrous Analysis, Working Examples, and Instructions for Making Apparatus. By Lieut.-Col. W. A. Ross, R.A. With 120 Illustrations. New Edition. Crown 8vo, 5s.

" The student who goes through the course of experimentation here laid down will gain a better insight into inorganic chemistry and mineralogy than if he had 'got up' any of the best text-books of the day, and passed any number of examinations in their contents."—*Chemical News.*

Commercial Chemical Analysis.

THE COMMERCIAL HANDBOOK OF CHEMICAL ANA-LYSIS; or, Practical Instructions for the determination of the Intrinsic or Commercial Value of Substances used in Manufactures, Trades, and the Arts. By A. NORMANDY. New Edition by H. M. NOAD, F.R.S. Cr. 8vo, 12s. 6d. cl.

" Essential to the analysts appointed under the new Act. The most recent results are given, and the work is well edited and carefully written."—*Nature.*

Brewing.
A HANDBOOK FOR YOUNG BREWERS. By HERBERT EDWARDS WRIGHT, B.A. An Entirely New Edition, much Enlarged.
[In the press.

Analysis and Valuation of Fuels.
FUELS: SOLID, LIQUID AND GASEOUS, Their *Analysis and Valuation.* For the Use of Chemists and Engineers. By H. J. PHILLIPS, F.C.S., Analytical and Consulting Chemist to the Great Eastern Railway. Crown 8vo, 3s. 6d. cloth. *[Just published*
" Ought to have its place in the laboratory of every metallurgical establishment, and wherever fuel is used on a large scale."—*Chemical News.*
" Mr. Phillips' new book cannot fail to be of wide interest, especially at the present time."—*Railway News.*

Dye-Wares and Colours.
THE MANUAL OF COLOURS AND · DYE-WARES : Their *Properties, Applications, Valuation, Impurities, and Sophistications.* For the use of Dyers, Printers, Drysalters, Brokers, &c. By J. W. SLATER. Second Edition, Revised and greatly Enlarged. Crown 8vo, 7s. 6d. cloth.
"A complete encyclopædia of the *materia tinctoria.* The information given respecting each article is full and precise, and the methods of determining the value of articles such as these, so liable to sophistication, are given with clearness, and are practical as well as valuable."—*Chemist and Druggist.*
" There is no other work which covers precisely the same ground. To students preparing or examinations in dyeing and printing it will prove exceedingly useful."—*Chemical News.*

Pigments.
THE ARTIST'S MANUAL OF PIGMENTS. Showing their Composition, Conditions of Permanency, Non-Permanency, and Adulterations; Effects in Combination with Each Other and with Vehicles; and the most Reliable Tests of Purity. Together with the Science and Arts Department's Examination Questions on Painting. By H. C. STANDAGE. Second Edition. Crown 8vo, 2s. 6d. cloth.
" This work is indeed *multum-in-parvo,* and we can, with good conscience, recommend it to all who come in contact with pigments, whether as makers, dealers or users."—*Chemical Review.*

Gauging. Tables and Rules for Revenue Officers, Brewers, etc.
A POCKET BOOK OF MENSURATION AND GAUGING : Containing Tables, Rules and Memoranda for Revenue Officers, Brewers, Spirit Merchants, &c. By J. B. MANT (Inland Revenue). Second Edition, Revised. Oblong 18mo, 4s. leather, with elastic band. *[Just published.*
" This handy and useful book is adapted to the requirements of the Inland Revenue Department, and will be a favourite book of reference. The range of subjects is comprehensive, and the arrangement simple and clear."—*Civilian.*
" Should be in the hands of every practical brewer."—*Brewers' Journal.*

AGRICULTURE, FARMING, GARDENING, etc.

Youatt and Burn's Complete Grazier.
THE COMPLETE GRAZIER, and FARMER'S and CATTLE-BREEDER'S ASSISTANT. A Compendium or Husbandry; especially in the departments connected with the Breeding, Rearing, Feeding, and General Management of Stock; the Management of the Dairy, &c. With Directions for the Culture and Management of Grass Land, the Management of Grain and Root Crops, the Arrangement of Farm Offices, the use of Implements and Machines, and on Draining, Irrigation, Warping, &c.; and the Application and Relative Value of Manures. By WILLIAM YOUATT, Esq., V.S., and ROBERT SCOTT BURN. A New Edition, partly Re-Written and greatly Enlarged by W. FREAM, B.Sc. Lond., LL.D. One large 8vo Volume, nearly 1,000 pages.
[In preparation.

Agricultural Facts and Figures.
NOTE-BOOK OF AGRICULTURAL FACTS AND FIGURES FOR FARMERS AND FARM STUDENTS. By PRIMROSE McCONNELL, Fellow of the Highland and Agricultural Society; late Professor of Agriculture, Glasgow Veterinary College. Third Edition. Royal 32mo, full roan, gilt edges, with elastic band, 4s.
" The most complete and comprehensive Note-book for Farmers and Farm Students that we have seen. It literally teems with information, and we can cordially recommend it to all connected with agriculture."—*North British Agriculturist.*

Flour Manufacture, Milling, etc.

FLOUR MANUFACTURE: A Treatise on Milling Science and Practice. By FRIEDRICH KICK, Imperial Regierungsrath, Professor of Mechanical Technology in the Imperial German Polytechnic Institute, Prague. Translated from the Second Enlarged and Revised Edition with Supplement. By H. H. P. POWLES, A.M.I.C.E. Nearly 400 pp. Illustrated with 28 Folding Plates, and 167 Woodcuts. Royal 8vo, 25s. cloth.

'This valuable work is, and will remain, the standard authority on the science of milling. . . The miller who has read and digested this work will have laid the foundation, so to speak, of a successful career; he will have acquired a number of general principles which he can proceed to apply. In this handsome volume we at last have the accepted text-book of modern milling in good, sound English, which has little, if any, trace of the German idiom."—*The Miller.*
"The appearance of this celebrated work in English is very opportune, and British millers will, we are sure, not be slow in availing themselves of its pages."—*Millers' Gazette.*

Small Farming.

SYSTEMATIC SMALL FARMING; or, *The Lessons of my Farm.* Being an Introduction to Modern Farm Practice for Small Farmers in the Culture of Crops; The Feeding of Cattle; The Management of the Dairy, Poultry and Pigs, &c. &c. By ROBERT SCOTT BURN, Author of "Outlines of Landed Estates' Management." Numerous Illusts., cr. 8vo, 6s. cloth.

"This is the completest book of its class we have seen, and one which every amateur farmer will read with pleasure and accept as a guide."—*Field.*
"The volume contains a vast amount of useful information. No branch of farming is left untouched, from the labour to be done to the results achieved. It may be safely recommended to all who think they will be in paradise when they buy or rent a three-acre farm."—*Glasgow Herald.*

Modern Farming.

OUTLINES OF MODERN FARMING. By R. SCOTT BURN. Soils, Manures, and Crops—Farming and Farming Economy—Cattle, Sheep, and Horses — Management of Dairy, Pigs and Poultry — Utilisation of Town-Sewage, Irrigation, &c. Sixth Edition. In One Vol., 1,250 pp., half-bound, profusely Illustrated, 12s.

"The aim of the author has been to make his work at once comprehensive and trustworthy, and in this aim he has succeeded to a degree which entitles him to much credit."—*Morning Advertiser.*   "No farmer should be without this book."—*Banbury Guardian.*

Agricultural Engineering.

FARM ENGINEERING, THE COMPLETE TEXT-BOOK OF. Comprising Draining and Embanking; Irrigation and Water Supply; Farm Roads, Fences, and Gates; Farm Buildings, their Arrangement and Construction, with Plans and Estimates; Barn Implements and Machines; Field Implements and Machines; Agricultural Surveying, Levelling, &c. By Prof. JOHN SCOTT, Editor of the "Farmers' Gazette," late Professor of Agriculture and Rural Economy at the Royal Agricultural College, Cirencester, &c. &c. In One Vol., 1,150 pages, half-bound, with over 600 Illustrations, 12s.

"Written with great care, as well as with knowledge and ability. The author has done his work well; we have found him a very trustworthy guide wherever we have tested his statements. The volume will be of great value to agricultural students."—*Mark Lane Express.*
"For a young agriculturist we know of no handy volume likely to be more usefully studied."—*Bell's Weekly Messenger.*

English Agriculture.

THE FIELDS OF GREAT BRITAIN: A Text-Book of Agriculture, adapted to the Syllabus of the Science and Art Department. For Elementary and Advanced Students. By HUGH CLEMENTS (Board of Trade). Second Ed., Revised, with Additions. 18mo, 2s. 6d. cl. [*Just published.*

"A most comprehensive volume, giving a mass of information."—*Agricultural Economist.*
"It is a long time since we have seen a book which has pleased us more, or which contains such a vast and useful fund of knowledge."—*Educational Times.*

Tables for Farmers, etc.

TABLES, MEMORANDA, AND CALCULATED RESULTS for Farmers, Graziers, Agricultural Students, Surveyors, Land Agents Auctioneers, etc. With a New System of Farm Book-keeping. Selected and Arranged by SIDNEY FRANCIS. Second Edition, Revised. 272 pp., waistcoat-pocket size, 1s. 6d. limp leather. [*Just published.*

"Weighing less than 1 oz., and occupying no more space than a match box, it contains a mass of facts and calculations which has never before, in such handy form, been obtainable. Every operation on the farm is dealt with. The work may be taken as thoroughly accurate, the whole of the tables having been revised by Dr. Fream. We cordially recommend it."—*Bell's Weekly Messenger.*
"A marvellous little book. . . . The agriculturist who possesses himself of it will not be disappointed with his investment."—*The Farm.*

Farm and Estate Book-keeping.

BOOK-KEEPING FOR FARMERS & ESTATE OWNERS.
A Practical Treatise, presenting, in Three Plans, a System adapted for all Classes of Farms. By JOHNSON M. WOODMAN, Chartered Accountant. Second Edition, Revised. Cr. 8vo, 3s. 6 l. cl. bds. ; or 2s. 6d. cl. limp. [*Just published.*
"The volume is a capital study of a most important subject."—*Agricultural Gazette.*
"Will be found of great assistance by those who intend to commence a system of book-keeping, the author's examples being clear and explicit, and his explanations, while full and accurate, being to a large extent free from technicalities."—*Live Stock Journal.*

Farm Account Book.

WOODMAN'S YEARLY FARM ACCOUNT BOOK. Giving a Weekly Labour Account and Diary, and showing the Income and Expenditure under each Department of Crops, Live Stock, Dairy, &c. &c. With Valuation, Profit and Loss Account, and Balance Sheet at the end of the Year, and an Appendix of Forms. Ruled and Headed for Entering a Complete Record of the Farming Operations. By JOHNSON M. WOODMAN, Chartered Accountant, Author of "Book-keeping for Farmers." Folio, 7s. 6d. half bound. [*culture.*
"Contains every requisite form for keeping farm accounts readily and accurately."—*Agri-*

Early Fruits, Flowers and Vegetables.

THE FORCING GARDEN ; or, How to Grow Early Fruits, Flowers, and Vegetables. With Plans and Estimates for Building Glasshouses, Pits and Frames. Containing also Original Plans for Double Glazing, a New Method of Growing the Gooseberry under Glass, &c. &c., and on Ventilation, Protecting Vine Borders, &c. With Illustrations. By SAMUEL WOOD. Crown 8vo, 3s. 6d. cloth.
"A good book, and fairly fills a place that was in some degree vacant. The book is written with great care, and contains a great deal of valuable teaching."—*Gardeners' Magazine.*
"Mr. Wood's book is an original and exhaustive answer to the question 'How to Grow Early Fruits, Flowers and Vegetables?'"—*Land and Water.*

Good Gardening.

A PLAIN GUIDE TO GOOD GARDENING ; or, How to Grow Vegetables, Fruits, and Flowers. With Practical Notes on Soils, Manures, Seeds, Planting, Laying-out of Gardens and Grounds, &c. By S. WOOD. Fourth Edition, with considerable Additions, &c., and numerous Illustrations. Crown 8vo, 3s. 6d. cloth.
"A very good book, and one to be highly recommended as a practical guide. The practical directions are excellent."—*Athenæum.*
"May be recommended to young gardeners, cottagers, and specially to amateurs, for the plain, simple, and trustworthy information it gives on common matters too often neglected."—*Gardeners' Chronicle.*

Gainful Gardening.

MULTUM-IN-PARVO GARDENING ; cr, How to make One Acre of Land produce £620 a-year by the Cultivation of Fruits and Vegetables ; also, How to Grow Flowers in Three Glass Houses, so as to realise £176 per annum clear Profit. By SAMUEL WOOD, Author of "Good Gardening," &c. Fifth and cheaper Edition, Revised, with Additions. Crown 8vo, 1s. sewed.
"We are bound to recommend it as not only suited to the case of the amateur and gentleman's gardener, but to the market grower."—*Gardeners' Magazine.*

Gardening for Ladies.

THE LADIES' MULTUM-IN-PARVO FLOWER GARDEN, and Amateurs' Complete Guide. By S. WOOD. With Illusts. Cr. 8vo, 3s. 6d. cl.
"This volume contains a good deal of sound, common sense instruction."—*Florist.*
"Full of shrewd hints and useful instructions, based on a lifetime of experience."—*Scotsman.*

Receipts for Gardeners.

GARDEN RECEIPTS. Edited by CHARLES W. QUIN. 12mo, 1s. 6d. cloth limp.
"A useful and handy book, containing a good deal of valuable information."—*Athenæum.*

Market Gardening.

MARKET AND KITCHEN GARDENING. By Contributors to "The Garden." Compiled by C. W. SHAW, late Editor of "Gardening Illustrated." 12mo, 3s. 6d. cloth boards. [*Just published.*
"The most valuable compendium of kitchen and market-garden work published."—*Farmer.*

Cottage Gardening.

COTTAGE GARDENING ; or, Flowers, Fruits, and Vegetables for Small Gardens. By E. HOBDAY. 12mo, 1s. 6d. cloth limp.
"Contains much useful information at a small charge."—*Glasgow Herald.*

LAND AND ESTATE MANAGEMENT, LAW, etc.

Hudson's Land Valuer's Pocket-Book.

THE LAND VALUER'S BEST ASSISTANT: Being Tables on a very much Improved Plan, for Calculating the Value of Estates. With Tables for reducing Scotch, Irish, and Provincial Customary Acres to Statute Measure, &c. By R. HUDSON, C.E. New Edition. Royal 32mo, leather, elastic band, 4s.

"This new edition includes tables for ascertaining the value of leases for any term of years; and for showing how to lay out plots of ground of certain acres in forms, square, round, &c., with valuable rules for ascertaining the probable worth of standing timber to any amount; and is of incalculable value to the country gentleman and professional man."—*Farmers' Journal.*

Ewart's Land Improver's Pocket-Book.

THE LAND IMPROVER'S POCKET-BOOK OF FORMULÆ, TABLES and MEMORANDA required in any Computation relating to the Permanent Improvement of Landed Property. By JOHN EWART, Land Surveyor and Agricultural Engineer. Second Edition, Revised. Royal 32mo, oblong, leather, gilt edges, with elastic band, 4s.'

"A compendious and handy little volume."—*Spectator.*

Complete Agricultural Surveyor's Pocket-Book.

THE LAND VALUER'S AND LAND IMPROVER'S COM-PLETE POCKET-BOOK. Consisting of the above Two Works bound to-gether. Leather, gilt edges, with strap, 7s. 6d.

"Hudson's book is the best ready-reckoner on matters relating to the valuation of land and crops, and its combination with Mr. Ewart's work greatly enhances the value and usefulness of the latter-mentioned. . . . It is most useful as a manual for reference."—*North of England Farmer.*

Auctioneer's Assistant.

THE APPRAISER, AUCTIONEER, BROKER, HOUSE AND ESTATE AGENT AND VALUER'S POCKET ASSISTANT, for the Valua-tion for Purchase, Sale, or Renewal of Leases, Annuities and Reversions, and of property generally; with Prices for Inventories, &c. By JOHN WHEELER, Valuer, &c. Fifth Edition, re-written and greatly extended by C. NORRIS, Surveyor, Valuer, &c. Royal 32mo, 5s. cloth.

"A neat and concise book of reference, containing an admirable and clearly-arranged list of prices for inventories, and a very practical guide to determine the value of furniture, &c."—*Standard.*

"Contains a large quantity of varied and useful information as to the valuation for purchase, sale, or renewal of leases, annuities and reversions, and of property generally, with prices for inventories, and a guide to determine the value of interior fittings and other effects."—*Builder.*

Auctioneering.

AUCTIONEERS: Their Duties and Liabilities. A Manual of Instruction and Counsel for the Young Auctioneer. By ROBERT SQUIBBS, Auctioneer. Second Edition, Revised and partly Re-written. Demy 8vo, 12s. 6d. cloth. [*Just published.*

"The position and duties of auctioneers treated compendiously and clearly."—*Builder.*

"Every auctioneer ought to possess a copy of this excellent work."—*Ironmonger.*

"Of great value to the profession. . . . We readily welcome this book from the fact that it treats the subject in a manner somewhat new to the profession."—*Estates Gazette.*

Legal Guide for Pawnbrokers.

THE PAWNBROKERS', FACTORS' AND MERCHANTS' GUIDE TO THE LAW OF LOANS AND PLEDGES. With the Statutes and a Digest of Cases on Rights and Liabilities, Civil and Criminal, as to Loans and Pledges of Goods, Debentures, Mercantile and other Se-curities. By H. C. FOLKARD, Esq., Barrister-at-Law, Author of "The Law of Slander and Libel," &c. With Additions and Corrections. Fcap. 8vo, 3s. 6d. cloth.

"This work contains simply everything that requires to be known concerning the department of the law of which it treats. We can safely commend the book as unique and very nearly perfect."—*Iron.*

"The task undertaken by Mr. Folkard has been very satisfactorily performed. . . . Such ex-planations as are needful have been supplied with great clearness and with due regard to brevity."—*City Press.*

Law of Patents.

PATENTS FOR INVENTIONS, AND HOW TO PROCURE THEM. Compiled for the Use of Inventors, Patentees and others. By G. G. M. HARDINGHAM, Assoc.Mem.Inst.C.E., &c. Demy 8vo, cloth, price 2s. 6d. [Just published.

Metropolitan Rating Appeals.

REPORTS OF APPEALS HEARD BEFORE THE COURT OF GENERAL ASSESSMENT SESSIONS, from the Year 1871 to 1885. By EDWARD RYDE and ARTHUR LYON RYDE. Fourth Edition, brought down to the Present Date, with an Introduction to the Valuation (Metropolis) Act, 1869, and an Appendix by WALTER C. RYDE, of the Inner Temple, Barrister-at-Law. 8vo, 16s. cloth,

" A useful work, occupying a place mid-way between a handbook for a lawyer and a guide to the surveyor. It is compiled by a gentleman eminent in his profession as a land agent, whose specialty, it is acknowledged, lies in the direction of assessing property for rating purposes."—*Land Agents' Record.*

" It is an indispensable work of reference for all engaged in assessment business."—*Journa. of Gas Lighting.*

House Property.

HANDBOOK OF HOUSE PROPERTY. *A Popular and Practical Guide to the Purchase, Mortgage, Tenancy, and Compulsory Sale of Houses and Land,* including the Law of Dilapidations and Fixtures; with Examples of all kinds of Valuations, Useful Information on Building, and Suggestive Elucidations of Fine Art. By E. L. TARBUCK, Architect and Surveyor. Fourth Edition, Enlarged. 12mo, 5s. cloth.

" The advice is thoroughly practical."—*Law Journal.*
" For all who have dealings with house property, this is an indispensable guide."—*Decoration.*
" Carefully brought up to date, and much improved by the addition of a division on fine art.
" A well-written and thoughtful work."—*Land Agent's Record.*

Inwood's Estate Tables.

TABLES FOR THE PURCHASING OF ESTATES, *Freehold, Copyhold, or Leasehold; Annuities, Advowsons, etc.,* and for the Renewing of Leases held under Cathedral Churches, Colleges, or other Corporate bodies, for Terms of Years certain, also for Lives; also for Valuing Reversionary Estates, Deferred Annuities, Next Presentations, &c.; together with SMART'S Five Tables of Compound Interest, and an Extension of the same to Lower and Intermediate Rates. By W. INWOOD. 23rd Edition, with considerable Additions, and new and valuable Tables of Logarithms for the more Difficult Computations of the Interest of Money, Discount, Annuities, &c., by M. FEDOR THOMAN, of the Société Crédit Mobilier of Paris. Crown 8vo, 8s. cloth.

" Those interested in the purchase and sale of estates, and in the adjustment of compensation cases, as well as in transactions in annuities, life insurances, &c., will find the present edition of eminent service."—*Engineering.*

" 'Inwood's Tables' still maintain a most enviable reputation. The new issue has been enriched by large additional contributions by M. Fedor Thoman, whose carefully arranged Tables cannot fail to be of the utmost utility."—*Mining Journal.*

Agricultural and Tenant-Right Valuation.

THE AGRICULTURAL AND TENANT-RIGHT-VALUER'S ASSISTANT. A Practical Handbook on Measuring and Estimating the Contents, Weights and Values of Agricultural Produce and Timber, the Values of Estates and Agricultural Labour, Forms of Tenant-Right-Valuations, Scales of Compensation under the Agricultural Holdings Act, 1883, &c. &c. By TOM BRIGHT, Agricultural Surveyor. Crown 8vo, 3s. 6d. cloth.

" Full of tables and examples in connection with the valuation of tenant-right, estates, labour, contents, and weights of timber, and farm produce of all kinds."—*Agricultural Gazette.*
" An eminently practical handbook, full of practical tables and data of undoubted interest and value to surveyors and auctioneers in preparing valuations of all kinds."—*Farmer.*

Plantations and Underwoods.

POLE PLANTATIONS AND UNDERWOODS: A Practical Handbook on Estimating the Cost of Forming, Renovating, Improving and Grubbing Plantations and Underwoods, their Valuation for Purposes of Transfer, Rental, Sale or Assessment. By TOM BRIGHT, F.S.Sc., Author of " The Agricultural and Tenant-Right-Valuer's Assistant," &c. Crown 8vo, 3s. 6d. cloth. [Just published.

" Will be found very useful to those who are actually engaged in managing wood."—*Bell's Weekly Messenger.*
" To valuers, foresters and agents it will be a welcome aid."—*North British Agriculturist.*
" Well calculated to assist the valuer in the discharge of his duties, and of undoubted interest and use both to surveyors and auctioneers in preparing valuations of all kinds."—*Kent Herald.*

A Complete Epitome of the Laws of this Country.

EVERY MAN'S OWN LAWYER: A Handy-Book of the Principles of Law and Equity. By A BARRISTER. Twenty-eighth Edition. Revised and Enlarged. Including the Legislation of 1890, and including careful digests of *The Bankruptcy Act*, 1890; the *Directors' Liability Act*, 1890; the *Partnership Act*, 1890; the *Intestates' Estates Act*, 1890; the *Settled Land Act*, 1890; the *Housing of the Working Classes Act*, 1890; the *Infectious Disease (Prevention) Act*, 1890; the *Allotments Act*, 1890; the *Tenants' Compensation Act*, 1890; and the *Trustees' Appointment Act*, 1890; while other new Acts have been duly noted. Crown 8vo, 688 pp., price 6s. 8d. (saved at every consultation!), strongly bound in cloth. [*Just published.*

. THE BOOK WILL BE FOUND TO COMPRISE (AMONGST OTHER MATTER)—

THE RIGHTS AND WRONGS OF INDIVIDUALS—LANDLORD AND TENANT—VENDORS AND PURCHASERS—PARTNERS AND AGENTS—COMPANIES AND ASSOCIATIONS—MASTERS, SERVANTS AND WORKMEN—LEASES AND MORTGAGES—CHURCH AND CLERGY, RITUAL—LIBEL AND SLANDER—CONTRACTS AND AGREEMENTS—BONDS AND BILLS OF SALE—CHEQUES, BILLS AND NOTES—RAILWAY AND SHIPPING LAW—BANKRUPTCY AND INSURANCE—BORROWERS, LENDERS AND SURETIES—CRIMINAL LAW—PARLIAMENTARY ELECTIONS—COUNTY COUNCILS—MUNICIPAL CORPORATIONS—PARISH LAW, CHURCH-WARDENS, ETC.—INSANITARY DWELLINGS AND AREAS—PUBLIC HEALTH AND NUISANCES—FRIENDLY AND BUILDING SOCIETIES—COPYRIGHT AND PATENTS—TRADE MARKS AND DESIGNS—HUSBAND AND WIFE, DIVORCE, ETC.—TRUSTEES AND EXECUTORS—GUARDIAN AND WARD, INFANTS, ETC.—GAME LAWS AND SPORTING—HORSES, HORSE-DEALING AND DOGS—INNKEEPERS, LICENSING, ETC.—FORMS OF WILLS, AGREEMENTS, ETC. ETC.

NOTE.—*The object of this work is to enable those who consult it to help themselves to the law; and thereby to dispense, as far as possible, with professional assistance and advice. There are many wrongs and grievances which persons submit to from time to time through not knowing how or where to apply for redress; and many persons have as great a dread of a lawyer's office as of a lion's den. With this book at hand it is believed that many a SIX-AND-EIGHTPENCE may be saved; many a wrong redressed; many a right reclaimed; many a law suit avoided; and many an evil abated. The work has established itself as the standard legal adviser of all classes, and also made a reputation for itself as a useful book of reference for lawyers residing at a distance from law libraries, who are glad to have at hand a work embodying recent decisions and enactments.*

. OPINIONS OF THE PRESS.

" It is a complete code of English Law, written in plain language, which all can understand. Should be in the hands of every business man, and all who wish to abolish lawyers' bills."—*Weekly Times.*

A useful and concise epitome of the law, compiled with considerable care."—*Law Magazine.*

"A complete digest of the most useful facts which constitute English law."—*Globe.*

" This excellent handbook. . . . Admirably done, admirably arranged, and admirably cheap."—*Leeds Mercury.*

' A concise, cheap and complete epitome of the English law. So plainly written that he who runs may read, and he who reads may understand."—*Figaro.*

" A dictionary of legal facts well put together. The book is a very useful one."—*Spectator.*

" A work which has long been wanted, which is thoroughly well done, and which we most cordially recommend."—*Sunday Times.*

" The latest edition of this popular book ought to be in every business establishment, and on every library table."—*Sheffield Post.*

Private Bill Legislation and Provisional Orders.

HANDBOOK FOR THE USE OF SOLICITORS AND ENGINEERS Engaged in Promoting Private Acts of Parliament and Provisional Orders, for the Authorization of Railways, Tramways, Works for the Supply of Gas and Water, and other undertakings of a like character. By L. LIVINGSTON MACASSEY, of the Middle Temple, Barrister-at-Law, and Member of the Institution of Civil Engineers; Author of " Hints on Water Supply." Demy 8vo, 950 pp., price 25s. cloth.

" The volume is a desideratum on a subject which can be only acquired by practical experience, and the order of procedure in Private Bill Legislation and Provisional Orders is followed. The author's suggestions and notes will be found of great value to engineers and others professionally engaged in this class of practice."—*Building News.*

" The author's double experience as an engineer and barrister has eminently qualified him for the task, and enabled him to approach the subject alike from an engineering and legal point of view. The volume will be found a great help both to engineers and lawyers engaged in promoting Private Acts of Parliament and Provisional Orders."—*Local Government Chronicle.*

OGDEN, SMALE AND CO. LIMITED, PRINTERS, GREAT SAFFRON HILL, E.C.